# STUDENT TECHNOLO

**HOLT, RINEHART AND WINSTON**

A Harcourt Classroom Education Company

**Austin** • New York • Orlando • Atlanta • San Francisco • Boston • Dallas • Toronto • London

# To the Teacher

*Student Technology Guide* contains one- to two-page blackline masters for selected lessons from *Geometry*. These masters provide computer and calculator activites that offer additional practice and alternative technology to that provided in *Geometry*. The technologies emphasized in these masters are geometry graphics software, graphics calculators, and computer spreadsheets.

Copyright © by Holt, Rinehart and Winston

All rights reserved. No part of this publication may be reproduced or transmitted in any form or by any means, electronic or mechanical, including photocopy, recording, or any information storage and retrieval system, without permission in writing from the publisher.

Teachers using GEOMETRY may photocopy complete pages in sufficient quantities for classroom use only and not for resale.

**Photo Credit**
Front Cover: Dale Sanders/Masterfile

Printed in the United States of America

ISBN 0-03-054337-1

5  6  7  054  03

# Table of Contents

| | | |
|---|---|---|
| **Chapter 1** | Exploring Geometry | **1** |
| **Chapter 2** | Reasoning in Geometry | **10** |
| **Chapter 3** | Parallels and Polygons | **14** |
| **Chapter 4** | Triangle Congruence | **20** |
| **Chapter 5** | Perimeter and Area | **28** |
| **Chapter 6** | Shapes in Space | **37** |
| **Chapter 7** | Surface Area and Volume | **41** |
| **Chapter 8** | Similar Shapes | **46** |
| **Chapter 9** | Circles | **53** |
| **Chapter 10** | Trigonometry | **62** |
| **Chapter 11** | Taxicabs, Fractals, and More | **69** |
| **Chapter 12** | A Closer Look at Proof and Logic | **75** |
| **Answers** | | **78** |

NAME _____ CLASS _____ DATE _____

# Student Technology Guide
## 1.1 The Building Blocks of Geometry, page 1

Geometry software is a constructive tool; that is, you can use it to make geometric objects from primitive objects such as points, line segments, rays, lines, and circles. The illustration at right shows some of the tools available in a typical toolbar, which will help you construct simple objects in the examples below.

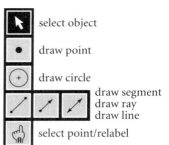

**Draw a triangle.**

Select the line segment tool. Place the mouse pointer in the sketch. Click anywhere in the sketch to create the first vertex, and drag the pointer to the desired location for the second vertex. Release. Click and drag from the second vertex to the desired location for the third. Release. Click and drag from the third vertex back to the first. Release.

**Draw a rectangle.**

Proceed in the same way described for making a triangle, but use four points this time. To display labels, place the mouse pointer on one vertex. Click once. Hold down SHIFT. Click once on each remaining vertex. From the main menu, select Display Show label.

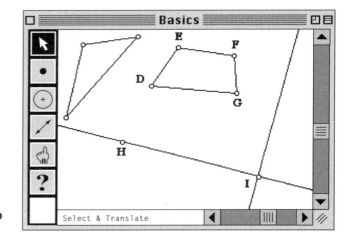

Drag point *E* so that it is directly above point *D*. Drag point *G* so that it is directly to the right of *D*. Drag point *F* so that it is directly above *G*. The reshaped figure is a rectangle.

**Draw two perpendicular lines.**

Drag across the toolbar to select the line tool. Click and drag from one location in the sketch to another. Click once. Release to see the line. Select the line with the pointer. Press SHIFT. Select a point on the line. From the main menu, select Construct Perpendicular line.

**Draw a line and a line parallel to it at a point not on the line.**

Create a line by using the same process as described above for perpendicular lines. Select the point tool and click at a location not on the line. Select the line with the pointer. Press SHIFT. Select the point not on the line. From the main menu, select Construct Parallel line.

Geometry                              Student Technology Guide        1

NAME _____  CLASS _____  DATE _____

# Student Technology Guide
## 1.1 The Building Blocks of Geometry, page 2

Now that you have used geometry graphics software to construct some simple geometric figures, you can practice your skills below.

**Use geometry graphics software to create each sketch.**

- **Draw several figures in one sketch, or make a new sketch for each situation.**
- **You may want to print your sketches and place them in a portfolio, along with descriptions of how you made them.**

1. Model an angle: Draw a ray. Then draw a second ray with the same endpoint as the first ray.

2. Draw three lines intersecting in one point.

3. Draw three parallel lines: First draw a line. Then place two distinct points in the sketch with neither point on the line.

4. Draw two lines perpendicular to the same line.

5. Illustrate how the plane is divided into regions by sketching three lines whose points of intersection determine a triangular region.

6. Draw a four-sided figure. Draw a line segment that lies entirely inside the figure. Adjust the figure so that the segment's endpoints lie inside the figure but part of it lies outside the figure.

7. Use line segments to construct the word **MATHEMATICS**.

8. Draw a line. Construct two lines perpendicular to the original line. Construct a line parallel to the original line and passing through a point on one of the two perpendicular lines.

**For exercises 9–11, use the following information:**

**To find the length of a segment, select the segment. From the main menu, select `Measure` `Length`. To find the measure of an angle, select a point other than the vertex on one side of the angle, hold down `SHIFT`, select the vertex, and select a point on the other side of the angle. From the main menu, select `Measure` `Angle`. Release `SHIFT`.**

9. Draw an angle and display its measure. Adjust the angle so that its measure is 45°.

10. Draw a line segment and display its length. Adjust the line segment so that its length is 2 inches.

11. Draw a four-sided figure. Find and display the lengths of the sides. Adjust the vertices so that each of the sides is 2 inches long.

# Student Technology Guide
## 1.2 Measuring Length

According to the Segment Addition Postulate:

If point *R* is between points *P* and *Q*, then $PR + RQ = PQ$.

A powerful aspect of geometry graphics software is that it gives you an instantaneous reading of the length of a segment.

**Use geometry graphics software to model the Segment Addition Postulate in Exercises 1–3.**

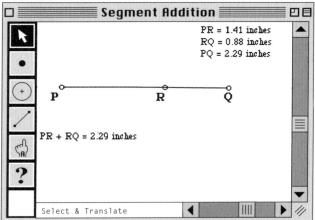

1. Draw a horizontal segment. Label the endpoints of the segment *P* and *R*. Then draw a second horizontal segment with point *R* as one of its endpoints. Label the other endpoint *Q*.

2. a. Use the measurement feature to find and display *PR*, *RQ*, and *PQ*.
   b. Use the calculation feature to find and display $PR + RQ$. To do this, first drag the pointer to create a box around the measurements. Select **Measure** **Calculate**, select the values you wish to add, and click **OK** to see the sum.

3. Drag point *R* closer to point *P*, then closer to point *Q*. (Be careful not to "bend" $\overline{PQ}$.) While you are doing this, observe what happens to the numbers that are displayed. Write a brief description of your observations. What geometric principle does this illustrate?

_____

4. Now drag point *R* so that you *do* bend $\overline{PQ}$. While you are doing this, observe what happens to the numbers that are displayed. Write a brief description of your observations. What geometric principle do you think this illustrates?

_____

5. Choose a measure greater than *PQ*. Drag point *R* to a location such that the sum $PR + RQ$ is equal to the measure you chose. How many such locations are there? Suppose you displayed a point at each of these locations. Make a conjecture about the type of figure the points would form.

_____

Geometry         Student Technology Guide

# Student Technology Guide
## 1.3 Measuring Angles

According to the Angle Addition Postulate:

If point S is in the interior of ∠PQR, then
m∠PQR = m∠PQS + m∠SQR.

Just as with the Segment Addition Postulate, you can use geometry graphics software to create a dynamic model of the Angle Addition Postulate.

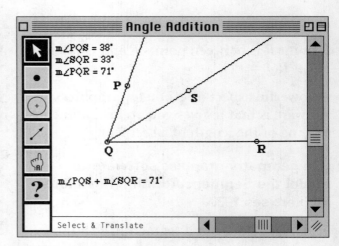

**Use geometry graphics software to model the Angle Addition Postulate in Exercises 1–3.**

1. Draw three rays with a common endpoint. Position them as shown above. Label them $\overrightarrow{QP}$, $\overrightarrow{QR}$, and $\overrightarrow{QS}$ as shown.

2. a. Use the measurement feature to find and display m∠PQS, m∠SQR, and m∠PQR.
   b. Use the calculation feature to find and display m∠PQS + m∠SQR.

3. Drag point S closer to point P, then closer to point R. While you are doing this, observe what happens to the numbers that are displayed. Write a brief description of your observations. What geometric principle does this illustrate?

   _____

4. Drag point S to a position in which it lies on the bisector of ∠PQR. Explain how you know that the position you chose is correct.

   _____

5. Notice that the Angle Addition Postulate specifies that point S must be in the *interior* of ∠PQR. Why do you think this is important? Explore what happens when you move point S to the exterior. Write a brief description of your observations.

   _____

6. Is the following statement true or false?

   If point S and point T are in the interior of ∠PQR, then
   m∠PQS + m∠SQT + m∠TQR = m∠PQR.

   Make a sketch to justify your answer.

   _____

**4**  Student Technology Guide  Geometry

NAME _____ CLASS _____ DATE _____

# Student Technology Guide
## 1.4 Exploring Geometry by Using Paper Folding

When you draw a geometric figure on a computer screen, it is impossible to actually fold the figure. However, most geometry graphics software has features that allow you to fold a figure *virtually*.

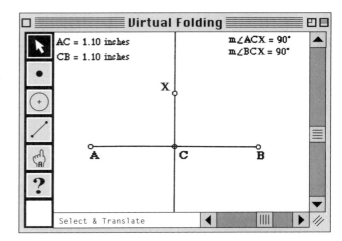

**"Fold" a line segment along its perpendicular bisector to complete Exercises 1 and 2.**

1. a. Draw a line segment. Label its endpoints *A* and *B*.
   b. Select $\overline{AB}$ and select **Construct** **Point at Midpoint**. Label this point *C*.
   c. Draw a line through point *C* perpendicular to $\overline{AB}$. The figure should look like the one above.
2. Devise and describe a method for "folding" $\overline{AB}$ into four equal parts.

---

**Using a new sketch, "fold" an angle along its bisector to complete Exercises 3 and 4.**

3. a. Draw two intersecting lines, $\overleftrightarrow{FG}$ and $\overleftrightarrow{HJ}$. Place a point where they intersect and label it *K*.
   b. Display the bisectors of $\angle GKJ$, $\angle JKF$, $\angle FKH$, and $\angle HKG$. Remember to select a point on one side of the angle, the vertex, a point on the other side, and then **Construct** **Angle bisector**.
4. Grab point *G* with the pointer and move it to several different positions. Use your observations to make a conjecture about the angle bisectors.

---

5. Make two sketches of the figure at right, first using the line tool and then using the parallel line and perpendicular line features.
6. In each sketch, drag *M* to several different positions. Then move *O*. Describe the difference between the behavior of the lines in each sketch.

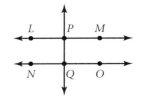

Geometry                                       Student Technology Guide    **5**

NAME _____ CLASS _____ DATE _____

# Student Technology Guide
## 1.5 Special Points in Triangles

In Lesson 1.5 of the textbook, you explored the behavior of altitudes, medians, perpendicular bisectors, and angle bisectors in triangles. Did you ever wonder how these special lines and segments behave in other figures—or whether they even exist in other figures? The exercises below explore these questions.

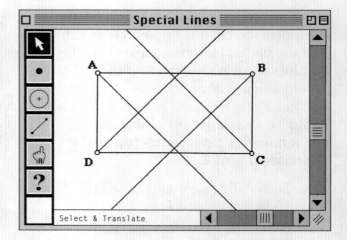

1. a. Draw four segments joined at their endpoints. Adjust their positions to form a rectangle like the one shown at right. Label the vertices A, B, C, and D, as shown.
   b. Bisect each angle. The resulting figure should look like the one shown.

2. Unlike the angle bisectors of a triangle, these angle bisectors do *not* intersect at one point. What is the relationship among the angle bisectors in this figure?

_____

_____

3. Drag $\overline{CD}$ closer to $\overline{AB}$, then farther away. Be careful that ABCD remains a rectangle. Do the angle bisectors ever intersect at one point? Use your observations to write a conjecture.

_____

4. Drag $\overline{CD}$ so that ABCD is no longer a rectangle. Do the angle bisectors ever intersect at one point? Use your observations to write a conjecture.

_____

5. Position $\overline{CD}$ so that ABCD is a rectangle again. Drag point B to different positions. Do the angle bisectors ever intersect at one point? Use your observations to write a conjecture.

_____

_____

6. Make a new sketch of rectangle ABCD. Use the midpoint and perpendicular line features to construct the four perpendicular bisectors. What can you say about them?

_____

**6**    Student Technology Guide    Geometry

NAME _____ CLASS _____ DATE _____

# Student Technology Guide
## 1.6 Motion in Geometry, page 1

Many designs and logos are easily created by copying and reflecting a basic figure. In the diagram at right, △ABO intersects lines *j* and *k* at point O.

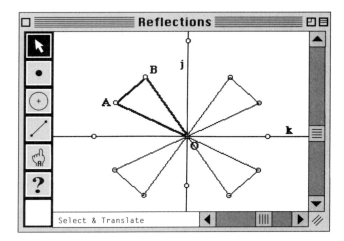

1. **a.** Draw vertical line *j* and horizontal line *k*.
   **b.** Draw △ABO with point O at the intersection of *j* and *k*.

2. **a.** Select line *j* with the pointer, and select **Transform** **Mark Mirror "j"**.
   Then use the pointer to box in △ABO, and select **Transform** **Reflect**.
   **b.** Using line *k* as the line of reflection, or *mirror*, reflect △ABO and the triangle from part **a.** Your diagram should contain four triangles like those shown above.

3. Explain why the four triangles have the same size and shape. _____

_____

4. Move line *k* up or down, or move line *j* to the left or right, maintaining their right angles. Do △ABO and the three triangles resulting from the reflection of △ABO still have the same size and shape? Explain.

   _____

   _____

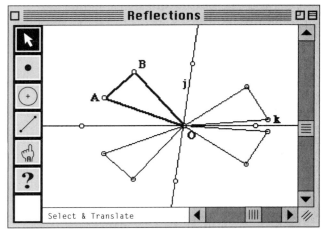

5. What happens if you change line *j* so that it is no longer vertical, or line *k* so that it is no longer horizontal?

   _____

   _____

   _____

**Geometry**                               **Student Technology Guide**

# Student Technology Guide
## 1.6 Motion in Geometry, page 2

Just as you can create a design by using a collection of reflections, you can generate a design by using a base figure and a set of rotations.

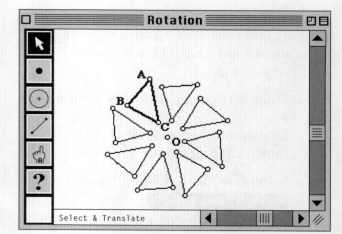

6. a. Draw △ABC, place point O as shown, and display labels for the points.
   b. Select point O, and select `Transform` `Mark Center "O"`. Box in △ABC with the pointer, and select `Transform` `Rotate...`. Choose 45°.
   c. Using the triangle you just generated, repeat part b.

7. What three reasons assure that the completed design is patterned the way it is?

_____

_____

Now you can explore what happens when you change the base figure. In the diagram at right, the base figure is rectangle ABCD and the center of rotation, point O, is the point where $\overline{AC}$ and $\overline{DB}$ intersect. A 45° rotation is used.

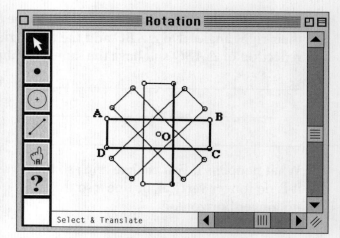

8. Draw rectangle ABCD. Find the center of rotation where the rectangle's diagonals intersect, label it, and perform successive rotations of 45° about that point. Describe your design.

_____

_____

8  Student Technology Guide                                    Geometry

NAME _____ CLASS _____ DATE _____

# Student Technology Guide
## 1.7 Motion in the Coordinate Plane

The rule $T(x, y) = (x + h, y + k)$, where $h$ and $k$ are fixed real numbers, gives a translation of $P(x, y)$ to $P'(x + h, y + k)$. If a geometric figure, such as a line, is defined by an equation, then you can use the translation rules to write and graph an equation for the geometric object after the translation is applied. Using a graphics calculator, you can explore lines and translations of them.

**Example:** The equation $y = 2x$ describes a line. Write an equation for the new line after you apply the rule $T(x, y) = (x + 3, y + (-2))$. Then graph both equations. How are they related?
- original equation: $y = 2x$
  new equation: $y = 2(x + 3) + (-2)$

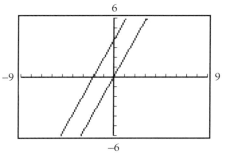

- Press [Y=]. For Y1 and Y2, press 2 [X,T,θ,n] [ENTER]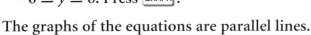
  2 [(] [X,T,θ,n] [+] 3 [)] [+] [(-)] 2 [ENTER].
- Press [WINDOW]. Enter the ranges $-9 \leq x \leq 9$ and $-6 \leq y \leq 6$. Press [GRAPH].

The graphs of the equations are parallel lines.

**Apply the translation $T(x, y) = (x + 3, y + (-2))$ to each equation. Write the new equation. Graph the original equation along with its translation on the same calculator display.**

1. $y = 2.5x$    2. $y = -2x$    3. $y = 4x$

_____   _____   _____

4. When you apply a translation to a given line, do a line and its image ever meet? Explain your response.

_____

_____

The rule $T(x, y) = (-x, -y)$ describes a transformation of $P(x, y)$ to $Q(-x, -y)$.

**In Exercises 5–8, apply the rule above to the given equation. Graph the original equation along with its image on the same calculator display.**

5. $y = 2.5x$    6. $y = -2x$    7. $y = -x$    8. $y = 4x$

9. Suppose that a line is described by $y = mx$. What can you say about the line that results when the rule $T(x, y) = (-x, -y)$ is applied? Justify your response.

_____

Geometry                                  Student Technology Guide

NAME _____ CLASS _____ DATE _____

# Student Technology Guide
## 2.1 An Introduction to Proofs

One day, Jamie made the statement that if you draw a triangle with lines through each vertex perpendicular to the side not containing that vertex, then the lines intersect inside the triangle. Michael decided to make this dynamic sketch to test the truth of the statement.

Is Jamie's claim true when other triangles are examined?

**Use geometry graphics software in each of the following exercises:**

1. a. Using the software, draw a triangle and show the labels for its vertices.
   b. Select a vertex and the side that does not contain it. Construct a perpendicular line. Display it as a heavy line by selecting the pointer, clicking once on the line, and selecting `Display` `Line Weight` `Thick`.
   c. Repeat part **b** for the other two vertices and sides. Label the point O where the heavy lines intersect.

2. Is point O inside the triangle you sketched? Drag a vertex to change the triangle's shape. Continue to drag vertices to view many different triangles and perpendicular lines. Does point O remain inside the triangle? What can you conclude?

   _____

**In Exercises 3–5, use geometry software to test the truth of each statement.**

3. Four points A, B, C, and D lie along the same horizontal line. Point B is to the right of A, C is to the right of B, and D is to the right of C. Then $AB + BC + CD = AD$. _____

4. Four points A, B, C, and D, lie in the same plane and $AB = BC = CD = DA$. The figure determined must be a square. _____

5. Four points A, B, C, and D, lie in the same plane. These points determine a four-sided figure. _____

6. If you believe that the statement in Exercise 3 is true, try to write a proof. If you believe that it is false, justify your position.

   _____

   _____

10  Student Technology Guide                           Geometry

NAME _____ CLASS _____ DATE _____

# Student Technology Guide
## 2.3 Definitions

The dynamic drawing at right shows ∠BAC and ∠DEF. The two angles were arbitrarily drawn in a plane.

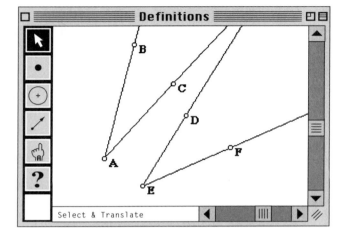

**In Exercises 1–4, use geometry software.**

1. Using the software, draw a diagram like the one shown.

2. a. Does the diagram illustrate the definition of a pair of adjacent angles? Explain your response.

   _____

   _____

   b. Explain how to modify the diagram so that the definition of adjacent angles is illustrated.

   _____

Rochelle invented a term called *nesting angles*. Here is her definition. *Two angles are nesting if they have the same vertex and the sides of one angle lie inside the space determined by the sides of the other angle.*

3. Modify the dynamic drawing you made in Exercise 1 to illustrate ∠DEF nesting inside ∠BAC. Describe your drawing in words.

   _____

Michael invented the term *opposing angles*. Here is his definition. *Two angles are opposing if the sides of one angle intersect the sides of the other angle.*

4. a. Modify the drawing you made in Exercise 1 to illustrate opposing angles ∠BAC and ∠DEF. Michael's definition allows for very different interpretations. Illustrate and describe two possible interpretations.

   _____

   b. Is Michael's definition of opposing angles a good definition? Answer the question knowing that Michael wants the word *opposing* to be meaningful.

   _____

   _____

Geometry                                                Student Technology Guide    **11**

NAME _____ CLASS _____ DATE _____

# Student Technology Guide
## 2.5 Conjectures That Lead to Theorems, page 1

Making a conjecture in geometry is similar to taking a sample and drawing a conclusion about a population in statistics. When you look for a conjecture, you select a representative sample from a population and attempt to see if what is true of the sample is true of the entire population.

Consider the population to be the set of all addition problems involving two even numbers. Consider the sample to be the addition problems below. In each of the five addition problems, all of the numbers being added are even.

2 + 2     4 + 12     28 + 36     72 + 100     12,428 + 54,844     1,345,002 + 18,334

Can a conclusion about sums of even numbers be drawn? To find out, begin by finding each of the sums above. Then use the following test: A number $n$ is even if $\frac{n}{2}$ is an integer. You can recognize an integer quotient by observing whether its decimal part is 0.

You can explore possible conclusions by using a graphics calculator.

- Using parentheses, enter each sum in the sample and divide by 2. Three of the quotients are shown on the display at right.

Each of the quotients is an integer. You have reason to believe that the sum of two even numbers is another even number.

- Test your conclusion by randomly creating some more addition problems involving only even numbers.

Again, each of the quotients is an integer. Thus, you have more reason to believe that the sum of two even numbers is another even number.

You can approach the same question about even-numbered sums by using a spreadsheet. The formula **2*INT(1000*RAND())** in columns A and B will give a random three-digit even number, and the formula **IF(A1+B1)/2=INT((A1+B1)/2),True,False)** in column C will test the sum to see if it is even.

**Use a spreadsheet to prove your conclusion about each population below.**

1. products of even numbers       2. sums of odd numbers

   _____                _____

3. squares of even numbers        4. squares of odd numbers

   _____                _____

12     Student Technology Guide     Geometry

NAME _____ CLASS _____ DATE _____

# Student Technology Guide
## 2.5 Conjectures That Lead to Theorems, page 2

After completing an activity about reflections across parallel lines, John and Melinda decided to see if they could draw a conclusion about the reflection of a figure across intersecting lines.

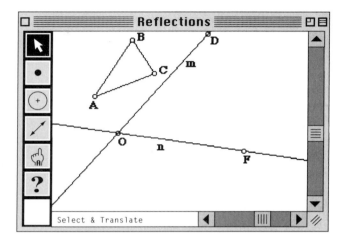

**Use geometry graphics software to explore conjectures about reflections of geometric figures across intersecting lines.**

5. a. Sketch △ABC.
   b. Sketch lines $m$ and $n$ intersecting at point $O$. (Do not sketch lines that cut through △ABC.)

6. Using the reflection tools with line $m$ as a mirror for △ABC, display the reflection of △ABC across line $m$. Then reflect across line $n$ the image that you just sketched of △ABC.

7. Let △A″B″C″ represent the final image after the reflection across line $n$. What observations can you make about the relationship between △ABC and △A″B″C″?

_____

John and Melinda know that a conjecture is not based on just one observation. Inductive reasoning requires you to examine many instances for a good sample.

8. Drag points $D$ and $F$ to make different lines of reflection. Observe the sketch. Modify △ABC so that it has different shapes. Do the observations you made in Exercise 7 still appear to be true? Explain.

_____

9. The sketch suggests that there is a rotation related to the double reflection of △ABC. Write a conjecture that you can reasonably infer from your exploration.

_____

_____

10. Is your conjecture from Exercise 9 still true if △ABC is placed in the interior of the angle formed by lines $m$ and $n$? Explain your response.

_____

Geometry                                    Student Technology Guide    **13**

NAME _____ CLASS _____ DATE _____

# Student Technology Guide
## 3.2 Properties of Quadrilaterals, page 1

The sketch at the right shows quadrilateral *ABCD* and the measures of its sides and its angles. Notice that the quadrilateral shown has no particularly noteworthy properties.

When you drag a vertex of quadrilateral *ABCD* to another location to form a different quadrilateral, you *continuously deform* the original figure.

Note: If you need to deform quadrilateral *ABCD* so that two opposite sides, say $\overline{AD}$ and $\overline{BC}$, are parallel, select *B* and $\overline{AD}$. Construct a parallel line through *B*. Then drag *C* so that it is on this line.

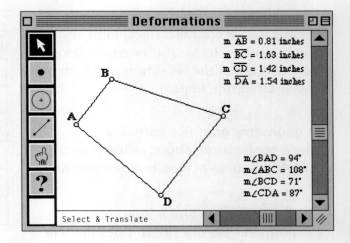

1. Sketch quadrilateral *ABCD*. Find and display the lengths of the sides and the measures of the angles.

**Use geometry graphics software to deform quadrilateral *ABCD* so that you know *by definition only* that it illustrates the specified figure. Explain your reasoning. On another paper, illustrate with a sketch.**

2. parallelogram _____

_____

3. rectangle _____

_____

4. rhombus _____

_____

5. square _____

_____

6. Matt wants to deform quadrilateral *ABCD* so that two sides are 1.5 inches long and the other two sides are 1.25 inches long. On another paper, sketch such a quadrilateral. Is the resulting figure unique?

_____

14    Student Technology Guide                                Geometry

NAME _____ CLASS _____ DATE _____

# Student Technology Guide
## 3.2 Properties of Quadrilaterals, page 2

The likelihood that quadrilateral *ABCD* sketched in Exercise 1 has no symmetry of any kind is very high. Now your job is to deform quadrilateral *ABCD* so that it becomes a symmetric quadrilateral.

Suppose that you want to deform quadrilateral *ABCD*, shown at right, giving it a line of symmetry such as line *n* in this diagram. You will need to begin by placing a point *Y* on, for example, $\overline{BC}$ and sketching line *n* through *Y* perpendicular to $\overline{BC}$.

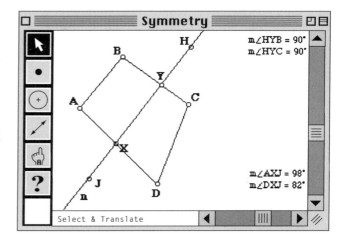

7. a. Return to the sketch of quadrilateral *ABCD* from Exercise 1. Delete everything in the sketch except the quadrilateral and the labels for its vertices. Modify the quadrilateral so that a line drawn perpendicular to one side will intersect the opposite side.
   b. Locate and mark point *Y* on $\overline{BC}$. Sketch line *n* perpendicular to $\overline{BC}$. Where *n* intersects $\overline{AD}$, mark and label point *X*. Find and display two points *H* and *J* on *n*, as shown above.
   c. Find and display m∠*AXJ* and m∠*DXJ* as shown.

8. a. Explain how to modify quadrilateral *ABCD* so that line *n* is a line of symmetry for $\overline{BC}$.

   _____

   b. Explain how to modify the diagram so that line *n* is a line of symmetry for $\overline{AD}$. Make sure that line *n* is still a line of symmetry for $\overline{BC}$.

   _____

   c. After completing part **b**, how do you know that quadrilateral *ABCD* now has line symmetry?

   _____

9. Return to the sketch of quadrilateral *ABCD* from Exercise 1. Delete everything in the sketch except the quadrilateral and the labels for its vertices. How would you modify quadrilateral *ABCD* so that it has rotational symmetry? What must be true of quadrilateral *ABCD* for it to have this symmetry?

   _____

Geometry                                   Student Technology Guide

NAME _____ CLASS _____ DATE _____

# Student Technology Guide
## *3.4* Proving That Lines Are Parallel

The sketch at right shows $\overleftrightarrow{AB}$ and $\overleftrightarrow{CD}$ with $\overleftrightarrow{XY}$ acting as a transversal. Notice that the lines cut by $\overleftrightarrow{XY}$ do not appear to be parallel.

Using facts about angles and lines cut by transversals, how can you adjust the sketch so that you know $\overleftrightarrow{AB}$ and $\overleftrightarrow{CD}$ will be parallel? To find out, answer the questions below.

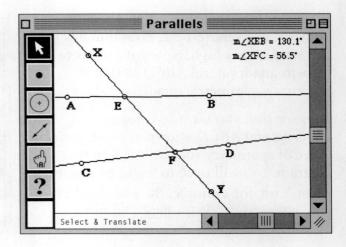

**Refer to the sketch at the right. Use geometry graphics software.**

1. a. Sketch two lines that do not appear to be parallel and a transversal for them.
   b. Display and label the points to match the sketch above.
   c. Find and display the measure of one angle, such as ∠XEB, where $\overleftrightarrow{XY}$ intersects $\overleftrightarrow{AB}$. In the sketch above, m∠XEB = 130.1°.

2. a. Choose an angle with vertex F where $\overleftrightarrow{XY}$ intersects $\overleftrightarrow{CD}$. In the sketch above, ∠XFC was selected and m∠XFC = 56.5°.
   b. Find and display the measure of the angle you chose in part **a**.
   c. Explain how you can adjust the measures of the two angles you chose to assure that $\overleftrightarrow{AB}$ and $\overleftrightarrow{CD}$ will be parallel. Then adjust $\overleftrightarrow{AB}$ or $\overleftrightarrow{CD}$ so that $\overleftrightarrow{AB} \parallel \overleftrightarrow{CD}$.

   _____

   _____

3. a. Repeat part **a** of Exercise 2 for a different angle.
   b. Explain how to adjust the angle from Exercise 1 and the angle from part **a** so that $\overleftrightarrow{AB}$ and $\overleftrightarrow{CD}$ will be parallel.

   _____

   _____

4. If the angle chosen in Exercise 1 has *E* as its vertex, explain why the angle chosen in Exercise 2 must have *F* as its vertex so that $\overleftrightarrow{AB}$ and $\overleftrightarrow{CD}$ will be parallel.

   _____

NAME _____ CLASS _____ DATE _____

# Student Technology Guide
## 3.5 The Triangle Sum Theorem

The Triangle Sum Theorem states that the sum of the measures of the angles in a triangle is always 180°. Using this and other geometric facts, you can write and solve an equation to find other angle measures.

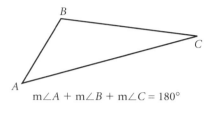

The example below shows how to use a calculator to find the measure of one angle in a triangle given the measures of the other two angles.

**Example:** In $\triangle ABC$, m$\angle A$ = 45.5° and m$\angle B$ = 72.6°. Find m$\angle C$. Let $a$, $b$, and $c$ be the measures of the angles. Then you can write $a + b + c = 180°$. Thus, $c = 180° - (a + b)$.

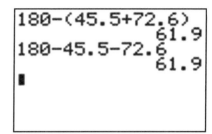

• Calculate $180° - (45.5° + 72.6°)$. Press 180 [ - ] [ ( ] 45.5 [ + ] 72.6 [ ) ] [ENTER].

• Alternatively, calculate $180° - 45.5° - 72.6°$.

From either calculation, m$\angle C$ = 61.9°.

**In Exercises 1 and 2, use the given information about $\triangle ABC$ to write an expression for the measure of the specified angle. Then use a calculator to find that measure.**

1. m$\angle A$ = 105.2° and m$\angle B$ = 42.7°; m$\angle C$

   a. _____

   b. _____

2. m$\angle B$ = 60.2° and m$\angle C$ = 59.4°; m$\angle A$

   a. _____

   b. _____

**In Exercises 3–7, write a calculator-ready numerical expression to find each angle measure. Then use a calculator to find it.**

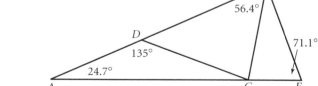

3. m$\angle BCA$ _____

4. m$\angle DCA$ _____

5. m$\angle BCD$ _____

6. m$\angle EBC$ _____

7. m$\angle BCE$ _____

Geometry           Student Technology Guide

NAME _____ CLASS _____ DATE _____

# Student Technology Guide
## 3.6 Angles in Polygons

From Lesson 3.6 of your textbook, you learned that the sum of the measures of the interior angles of a convex quadrilateral is 360°. As the example below illustrates, you can use algebra and a calculator to find unknown angles measures in quadrilaterals.

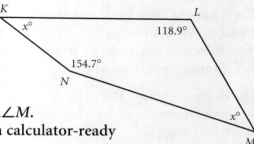

**Example:** Given quadrilateral KLMN, find m∠K and m∠M.
- Write an equation. Then transform it into a calculator-ready expression for $x$.

$$2x + (154.7 + 118.9) = 360 \longrightarrow x = \frac{360 - (154.7 + 118.9)}{2}$$

- Evaluate the expression by using the following key sequence: ( 360 − ( 154.7 + 118.9 ) ) ÷ 2 ENTER

Since $x = 43.2$, m∠K = m∠M = 43.2°.

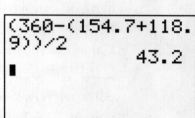

Notice in the example how parentheses are used:
- to group measures whose sum is subtracted from 360°
- to group the terms in the numerator

**For Exercises 1 and 2, write an equation to find $x$. Then transform it into a calculator-ready expression for $x$, and find $x$.**

1.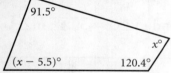

   a. _____
   b. _____

2. 
   a. _____
   b. _____

3. How many sides does a convex polygon have if the sum of the measures of its interior angles is 2400°?
   Debbie wrote $2400 = (n - 2)180$ and made the calculator evaluation shown at the right. Show that it is correct. Interpret the answer on the display. How do you know?

   ```
   2400/180+2
          15.3333333333
   ```

   _____
   _____
   _____

18   Student Technology Guide   Geometry

NAME _____ CLASS _____ DATE _____

# Student Technology Guide
## 3.7 Midsegments of Triangles and Trapezoids

When you carry out the procedure described below, you perform an *iteration*.
① Sketch △ABC as shown at right.
② Locate midpoints $M_1$ and $N_1$ of sides $\overline{AB}$ and $\overline{BC}$, respectively.
③ Construct segments $\overline{M_1B}$ and $\overline{N_1B}$.
④ Locate midpoints $M_2$ and $N_2$ of segments $\overline{M_1B}$ and $\overline{N_1B}$, respectively.
⑤ Continue creating segments and locating midpoints as described in Steps 3 and 4.

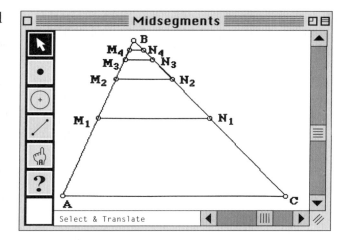

**Use geometry graphics software in each of the following exercises:**

1. Carry out the procedure described above to get a sketch like the one shown. Label the points as shown. _____

2. Find and display $AC$, $M_1N_1$, $M_2N_2$, $M_3N_3$, and $M_4N_4$. _____

3. Suppose that $M_kN_k$ is the $k$th segment drawn according to the process decsribed above. Write an equation for $M_kN_k$ in terms of $AC$. _____

4. Modify △ABC so that $AB = BC = AC$; that is, make △ABC equilateral.
   a. What can you say about $\triangle M_kN_kB$ for each positive integer $k$?
   b. How are the lengths of the sides of $\triangle M_kN_kB$ related to those of $\triangle M_{k+1}N_{k+1}B$? _____

5. Viewing the diagram above as a set of stacked trapezoids, locate and mark the midpoints of $\overline{AM_1}$ and $\overline{CN_1}$. Call the midpoints $R_1$ and $S_1$ respectively. Locate and mark the midpoints of $\overline{M_1M_2}$ and $\overline{N_1N_2}$, $\overline{M_2M_3}$ and $\overline{N_2N_3}$, and so on. Call these midpoints $R_2$ and $S_2$, $R_3$ and $S_3$, and so on, respectively. Find and display $R_1S_1$, $R_2S_2$, and so on.

6. Suppose that $\overline{R_kS_k}$ is the $k$th segment drawn according to the process decsribed in Exercise 5. Write an equation for $R_kS_k$ in terms of $AC$. _____

Geometry                                         Student Technology Guide    19

NAME _____ CLASS _____ DATE _____

# Student Technology Guide
## 4.3 Analyzing Triangle Congruence, page 1

Geometry software can be used to help model real-world situations and solve related problems. Consider the following problem:

An embankment makes a 20° angle with the horizontal. This is modeled by ∠CAB. Two vertical poles $\overline{EF}$ and $\overline{DG}$, each 10 feet high, are supported by cables anchored downslope. Each cable makes a 60° angle with the embankment. Are the cables the same length?

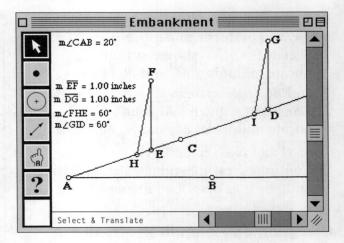

**In Exercises 1–3, use geometry graphics software to answer the question above.**

1. a. Sketch ∠CAB such that m∠CAB = 20°. Make $\overrightarrow{AB}$ horizontal.
   b. Place two points on $\overrightarrow{AC}$, label them E and D, and then sketch $\overline{EF}$ and $\overline{DG}$ such that they are vertical and have the same length. (In the sketch above, 1.00 inch represents 10 feet.)
   c. Place two points, H and I, on $\overrightarrow{AC}$, and then construct $\overline{HF}$ and $\overline{IG}$ such that m∠FHE = m∠GID = 60°.

2. Does the sketch suggest that HF = IG? Display HF and IG to confirm your observation. Then justify your response by using a triangle congruence theorem.

   _____

   _____

3. Briefly explain how an accurate computer-aided sketch helps you to be certain about the equality of the cable lengths.

   _____

   _____

4. Modify your sketch so that m∠CAB = 15°. Keep $\overrightarrow{AB}$ horizontal. Make adjustments so that parts **b** and **c** of Exercise 1 stay the same. Do the support cables still have the same length? What can you conclude about the given information?

   _____

   _____

NAME _____ CLASS _____ DATE _____

# Student Technology Guide
## 4.3 Analyzing Triangle Congruence, page 2

You can use computer modeling to explore navigation problems. When you plot courses, direction is measured from due north. The direction northeast means a path that makes a 45° angle clockwise from due north, represented by ∠NAB. The sketch also shows a southeasterly course, $\overline{YG}$. A southeasterly path is one that makes a 135° angle clockwise from due north, represented by ∠N'YG. Select **File** **New Sketch**.

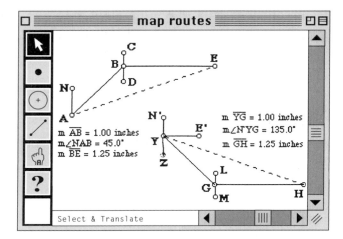

For the exercises below, use the following information: a person starting at point A travels northeast for 100 miles (represented by 1 inch) then due east for 125 miles (represented by 1.25 inches). A second traveler begins at point Y and travels southeast for 100 miles and due east for 125 miles.

5. a. Sketch ∠NAB such that m∠NAB = 45° and $\overline{AB}$ such that AB = 1. Then sketch a path, $\overline{BE}$, 1.25 inches due east. Sketch a short line segment, $\overline{CD}$, that contains B as shown.
   b. Find and display the measures of $\overline{AB}$, ∠NAB, and $\overline{BE}$.
   c. In similar fashion, sketch the rest of what is shown above. Make sure that your measurements for $\overline{YG}$, ∠N'YG, and $\overline{GH}$ agree with what is shown. Note also that $\overline{YE'}$ ' $\overline{YN'}$.

6. Use measurement features of the software to show that △ABE ≅ △YMH and therefore AE = YH.

7. Show that △ABE ≅ △YMH, so AE = YH. Use the space at right for your proof.

8. a. Use software to model each set of travel directions. Person A travels due east for 100 miles to point B and northeast for 200 miles, arriving at point C. Person X travels due west for 200 miles to point Y then southwest for 100 miles, arriving at point Z.
   b. Use the software and your reasoning to prove that AC = XZ. _____

Geometry                                    Student Technology Guide    21

# Student Technology Guide
## 4.5 Proving Quadrilateral Properties, page 1

Using only points and parallel lines, you can sketch a parallelogram. The sketch at right was made by following the steps below.

① Sketch $\overleftrightarrow{AB}$ as shown.

② Place point D in the plane but not on $\overleftrightarrow{AB}$.

③ Sketch $\overleftrightarrow{AD}$.

④ Construct a line through D parallel to $\overleftrightarrow{AB}$.

⑤ Construct a line through B parallel to $\overleftrightarrow{AD}$. Label C as the point where this line meets the line drawn in Step 4.

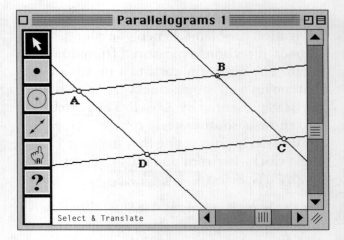

Once a parallelogram is drawn, you can observe properties of its sides, its angles, and its diagonals.

**Use geometry graphics software to confirm parallelogram properties.**

1. Create a parallelogram by following the process outlined above.

2. What would you do to demonstrate that opposite sides of any parallelogram have the same length?

3. How would you demonstrate that opposite angles of any parallelogram have the same measure?

4. A fellow student asks you to demonstrate a fact about the diagonals of a parallelogram. What fact would you demonstrate? How?

5. Another student asks for a fact about the triangles that result from drawing one diagonal. What would you tell that student? How would you use the software to demonstrate it?

NAME _____ CLASS _____ DATE _____

# Student Technology Guide
## 4.5 Proving Quadrilateral Properties, page 2

6. a. Return to the original sketch you made in Exercise 1. Find and display the lengths of the four sides. Adjust your sketch so that all four sides have the same length.
   b. One of your classmates sees you adjust your sketch and asks what type of quadrilateral you formed. What would your answer be?

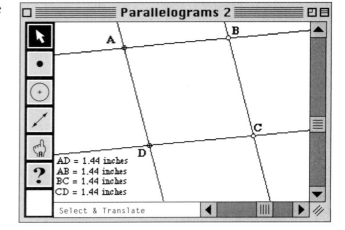

   _____

   c. Your classmate seems to recall a fact about the diagonals of the quadrilateral you identified, but cannot remember exactly what that fact is. What is true of the diagonals of such a quadrilateral? Explain how to use the software to demonstrate that fact.

   _____
   _____
   _____

7. A classmate argues as follows:
   > If you drag one or more vertices so that one angle of the quadrilateral measures 90° and you maintain the equality of the lengths of the four sides, the result will be a square.

   Do you agree with this claim? If so, how would you demonstrate its soundness by using the software and by using reason?

   _____
   _____
   _____

Recall that an *isosceles trapezoid* is a trapezoid that has exactly one pair of opposite sides congruent.

8. a. Make a new sketch of an isosceles trapezoid. Display the equality of the two congruent side lengths.
   b. How would you use the software to disprove the following claim?
      > The diagonals of an isosceles trapezoid bisect one another.

   _____
   _____

Geometry                                    Student Technology Guide    23

# Student Technology Guide
## 4.6 Conditions for Special Quadrilaterals, page 1

Shown at right is a sketch of convex quadrilateral *ABCD*. The diagonals' point of intersection is labeled *G*, and their midpoints have been located and labeled *E* and *F*.

In Lesson 4.6, you learned that when a certain set of facts are given, you can use those facts to tell what special quadrilateral is formed, if any. In the exercises below, you will use given facts about diagonals to identify the special nature of the quadrilateral.

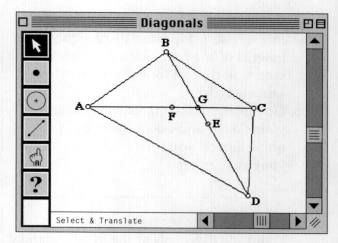

One statement about convex quadrilaterals that can be concluded from the sketch is the following:

If the midpoints of the diagonals are different points—that is, *E* and *F* and *G* are not all the same—then the quadrilateral is not a parallelogram. Therefore, it is not a rectangle, square, or rhombus.

**Use geometry graphics software as needed.**

1. Open a new sketch. Then make a sketch like the one above. Make sure *E*, *F*, and *G* are distinct. Drag point *E* so that it coincides with *G*, and make sure that point *F* does not coincide with points *G* and *E*. What can you say and what can you not say about quadrilateral *ABCD*?

   _____

   _____

2. Drag point *F* so that it coincides with *G* and *E*. What can you say about quadrilateral *ABCD*? What can you not say about quadrilateral *ABCD* until more information is given?

   _____

   _____

3. List the steps you would take in order to make quadrilateral *ABCD* a rhombus by making adjustments only to points *E* and *F* and to the diagonals. Then make those adjustments.

   _____

   _____

NAME _____ CLASS _____ DATE _____

# Student Technology Guide
## 4.6 Conditions for Special Quadrilaterals, page 2

Now that you have used geometry graphics software together with theorems about special quadrilaterals, you can apply your experience to making a classification chart for them. The result will be a bulletin-board display that contains a diagram like the one shown.

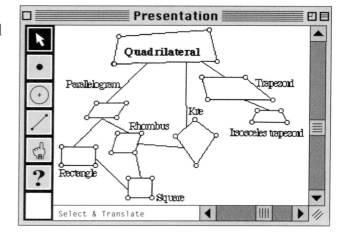

In this activity, you will learn how to add text to your sketch. Proceed as follows:

- Using [hand icon], click and drag to form a dashed box for the text.
- Type the text. When finished, click the mouse button once.
- With the pointer, drag the highlighted text to the location where you want it to appear as a caption.

**Use geometry graphics software as directed.**

4. a. Open a new sketch. Draw a quadrilateral as shown above. Use measurement tools to make sure that the quadrilateral does *not* illustrate any special quadrilateral properties. Enter the text *Quadrilateral*.
   b. Sketch a small parallelogram. Use measurement tools to make sure that the parallelogram is *not* a special parallelogram: a rectangle, a square, or a rhombus. Then hide everything you used to make the parallelogram except its sides and vertices. To do this, click once on the figure to be eliminated and select **Display** **Hide Segment**.
   c. Draw a line segment from the general quadrilateral to your parallelogram. Hide the endpoints of the line segment. Drag your parallelogram so that it is located as shown above. Enter the text *Parallelogram*.

5. Using the procedure in Exercise 4, add the following categories:
   - *Kite* and *Trapezoid*
   - *Rectangle, Rhombus,* and *Isosceles trapezoid*
   - *Square*
   Complete the set of links.

6. a. Check your diagram for accuracy and completeness.
   b. Write a brief summary of how you used theorems to make sure each figure represents its exact definition and *not* something more specialized.

_____

_____

Geometry

# Student Technology Guide
## 4.8 Constructing Transformations, page 1

You may be surprised to learn that you can construct a rhombus by using only one segment and reflections across lines. In the sketch at right, you see $\overline{AB}$ and line $k$ passing through $B$.

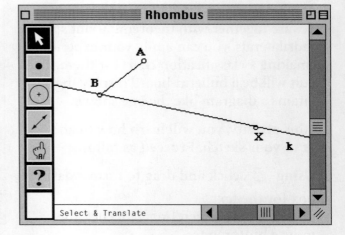

**Refer to the sketch at the right to start the construction of a rhombus.**

1. a. Make a sketch like the one shown. Label the segment and line as shown.
   b. Select line $k$ as the line of reflection and $\overline{AB}$ as the object to be reflected, and then reflect $\overline{AB}$ across line $k$. Label the reflection of point $A$ as point $C$. Explain how you know that $\overline{AB} \cong \overline{BC}$. _____

   c. After completing part **b**, you will have two adjacent sides of the rhombus left to be constructed by means of transformations. What would you do next to complete the rhombus with one or more reflections? Give your response below and carry out your strategy in your sketch.

2. a. The diagram at the right shows square $PQRS$ with line $k$ passing through points $P$ and $R$. Open a new sketch, and sketch $\overline{PQ}$ and line $k$ passing through $P$. Place point $X$ on $k$.
   b. In order to use line $k$ as the line of reflection so that the image of $\overline{PQ}$ will be perpendicular to $\overline{PQ}$, what should m∠$QPX$ equal? _____
   Adjust line $k$ so that m∠$QPX$ equals your response above.
   c. Reflect $\overline{PQ}$ across line $k$. You will have two adjacent sides of the square left to be constructed by means of transformations. How would you complete the square with one or more reflections? Write your strategy below and carry it out in your sketch.

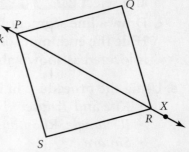

# Student Technology Guide
## 4.8 Constructing Transformations, page 2

In Exercise 2, you saw that you could construct a square from one line segment and a series of reflections by choosing to make the line of reflection and the segment form a special angle. The following question now arises:

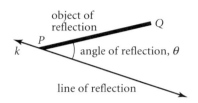

Is it possible to construct a regular polygon with $n$ sides from one line segment and a set of reflections?

To find out, consider the diagram at right. It shows angle $\theta$ (Greek letter *theta*) between $\overline{PQ}$ and line $k$. If $\theta$ has the right value, then it may be possible to carry out the construction described.

3. Suppose that you have a regular polygon with $n$ sides. Write an expression for one-half the measure of one interior angle. Then evaluate that expression for $n = 6$. _____

4. a. Using your software, open a blank sketch. Sketch $\overline{PQ}$ and line $k$ intersecting $\overline{PQ}$ at $P$. Find and label another point, $X$, on line $k$.
   b. Find and display m$\angle QPX$. Adjust $k$ so that m$\angle QPX$ equals the measure found in Exercise 3.
   c. Reflect $\overline{PQ}$ across line $k$. Label the image of $Q$ as $Q'$.
   d. Sketch a segment with $Q'$ as an endpoint. Place point $X'$ on it. Adjust the segment so that m$\angle PQ'X'$ equals the measure found in Exercise 3.
   e. Reflect $\overline{PQ}$ across $Q'X'$.
   f. Continue the process in parts **a–e** until a polygon is formed.

5. How do you know that the polygon formed in Exercise 4 is a regular hexagon?
   _____

You can also construct a regular polygon by using one line segment, its midpoint, and a set of rotations.

6. Open a new sketch. Sketch $\overline{AB}$ and label its endpoints. Label its midpoint as $O$. Using this sketch as an aid, describe how to construct a regular hexagon.

   _____
   _____
   _____
   _____

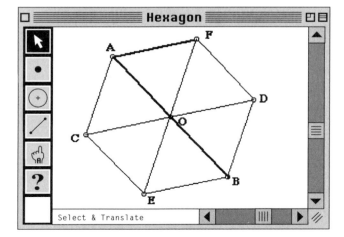

Geometry

NAME _____ CLASS _____ DATE _____

# Student Technology Guide
## 5.2 Areas of Triangles, Parallelograms, and Trapezoids

The example below shows you how to use a calculator to find the area of the trapezoid at right.

**Example:** Identify the needed information and set aside extraneous information.
height: 3.28 in.  bases: 4.56 in. and 6.72 in.
Using the needed information, apply the formula $A = \frac{1}{2}(b_1 + b_2)h$. (Note: Use 0.5 rather than $\frac{1}{2}$.)
Press .5 [×] [(] 6.72 [+] 4.56 [)] [×] 3.28 [ENTER].
Rounded to the nearest hundredth, the area equals 18.50 in.²

Sometimes you need to write and simplify an area expression before you use a calculator. The resulting key sequence is often simpler, and you will be less prone to error.

**Example:** To the nearest hundredth of a unit, find the area of the shaded region at right. A square is the center of four congruent rectangles. Simply find their total area by multiplying the area of one of them by 4. Press 4 [×] 3.84 [×] 2.24 [+] 2.24 [$x^2$] [ENTER].
To the nearest hundredth, the area equals 39.42 square units.

**Using a calculator, find the area of each shaded region. Write the simplest area expression. Give answers to the nearest hundredth of a square unit.**

1.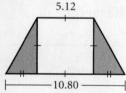
   5.12
   10.80

2.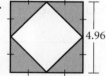
   4.96

3.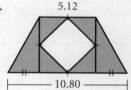
   5.12
   10.80

To find the length of one side of a square whose area equals 200 square units, press [2nd] [$x^2$] 200 [)] [ENTER]

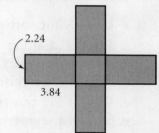

√(200)
   14.14213562

**To the nearest tenth, find the side length of a square with given area.**

4. 240 _____  5. 250 _____  6. 260 _____

28  Student Technology Guide  Geometry

# Student Technology Guide
## 5.3 Circumferences and Areas of Circles, page 1

Now that you have used the formulas for the areas of triangles, parallelograms, trapezoids, and circles, you are in a position to explore problems in which multiple figures are superimposed on one another.

The sketch at right shows square $ABCD$ with circle $E$ inscribed in it. Point $F$ is the midpoint of $\overline{BC}$. The equality $AB = AD$ and the four perpendicular lines ensure that $ABCD$ is a square.

**Use geometry graphics software as necessary.**

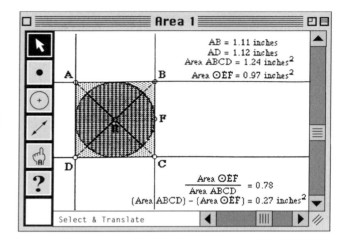

1. Prove that if quadrilateral $ABCD$ is a parallelogram, lines containing adjacent sides are perpendicular, and $AB = AD$, then quadrilateral $ABCD$ is a square. Use the space at the right for your paragraph proof.

2. a. Construct a square and label its vertices.
   b. Locate and label the midpoint of one side.
   c. Draw the diagonals and label their intersection.
   d. Using the point of intersection as the center and the midpoint as a point on a circle, draw the inscribed circle.
   e. Display each area by selecting the figure (you must highlight all four vertices of the square) and selecting **Measure** **Area**.

3. Use the pointer to box in the displayed areas, and select **Measure** **Calculate** to find the ratio of the circle's area to the square's and the difference of their areas. Drag points to create different dynamic sketches. What can you say about the measurements found and displayed?

_____

4. a. Let $s$ represent the side length of square $ABCD$. Write formulas for $\dfrac{\text{area circle } E}{\text{area square } ABCD}$ and (area of square $ABCD$) − (area of circle $E$) in terms of $s$. _____
   b. How do your formulas justify your answer to Exercise 3?

_____

Geometry                                              Student Technology Guide

NAME _____ CLASS _____ DATE _____

# Student Technology Guide
## 5.3 Circumferences and Areas of Circles, page 2

On the preceding page, you drew a circle inside a square. Now consider drawing a square inside a circle. This is shown at right.

Is it reasonable to suspect that the conclusions drawn in the preceding exercises remain true in this case?

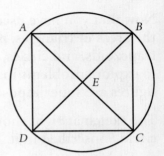

5. a. Write a conjecture about whether $\dfrac{\text{area of square } ABCD}{\text{area of circle } E}$ is fixed or varies.

   _____

   b. Write a conjecture about whether (area of circle $E$) − (area of square $ABCD$) is fixed or varies.

   _____

Now you can use geometry graphics software to provide verification of your conjectures from Exercise 5.

6. a. Construct square $ABCD$ and label its vertices. Calculate $AB$ and $AD$ to ensure that they are equal and that you have a square.
   b. Draw the diagonals and label their intersection $E$.
   c. Using $E$ as the center and one vertex of the square as a point on a circle, draw the circumscribed circle.
   d. Find and display the areas of square $ABCD$ and circle $E$.
   e. Find and display $\dfrac{\text{area of square } ABCD}{\text{area of circle } E}$ and (area of circle $E$) − (area of square $ABCD$).

7. Note what the display shows as the ratio and the difference. Drag points to make different squares inscribed in different circles. How do the sketches support the conjectures in Exercise 5?

   _____
   _____
   _____

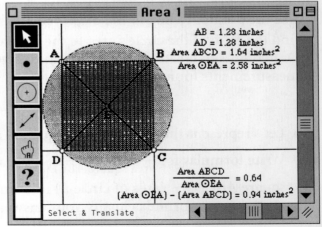

8. Use software to investigate $\dfrac{\text{perimeter of square } ABCD}{\text{circumference of circle } E}$ and (circumference of circle $E$) − (perimeter of square $ABCD$).

   _____
   _____

NAME _____ CLASS _____ DATE _____

# Student Technology Guide
## 5.4 The Pythagorean Theorem

To find a circle's radius given its area, or to find a right triangle's side length given the lengths of the other sides, you solve an equation by taking a square root. The examples show how to use a calculator to solve such problems.

**Example:**  a. In $\triangle ABC$, m$\angle C = 90°$, $AB = 24$, and $BC = 15$. To the nearest tenth, find $AC$.
b. The area of a circle is 300 square units. To the nearest tenth, find its radius.

```
√(24²−15²)
           18.734994
√(300/π)
           9.772050238
```

a. Write and solve an equation.
$(BC)^2 + (AC)^2 = (AB)^2 \rightarrow AC = \sqrt{(AB)^2 - (BC)^2}$
Press [2nd] [x²] 24 [x²] [−] 15 [x²] [ ) ] [ENTER]
To the nearest tenth of a unit $AC \approx 18.7$ units.

b. Write and solve an equation. $\pi r^2 = 300 \rightarrow r = \sqrt{\dfrac{300}{\pi}}$
Press [2nd] [x²] 300 [÷] [2nd] [^] [ ) ] [ENTER].
To the nearest tenth, the radius is about 9.8 units.

**Use a calculator to find each length. Give answers to the nearest tenth of a unit.**

1. In $\triangle KLM$, m$\angle L = 90°$, $KM = 200$, and $KL = 150$. Find $ML$. _____

2. In $\triangle XYZ$, m$\angle Y = 90°$, $XZ = 1234$, and $XY = 550$. Find $YZ$. _____

3. The area of a circle is 360 square units. Find its radius. _____

4. The area of a circle is $\pi^2$ square units. Find its radius. _____

To find the total distance $x + y$ in the diagram at right, you need to use the Pythagorean Theorem to enter the expression for $x$, press [+], and then use the theorem again to enter the expression for $y$. Finally, press [ENTER].

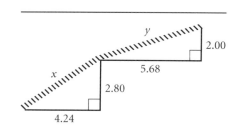

5. To the nearest tenth of a unit, find $x + y$. _____

6. a. Which is greater, $x$ or $y$? Modify the expression for $x + y$ by changing it to $x - y$. What does the calculator display tell you about which quantity is larger? _____

   b. By how much is one quantity greater than the other? _____

7. a. Suppose that the diagram is changed so that all four given lengths are 4.50. Write a numerical expression for $x + y$. _____

   b. Use a calculator and your expression from part a to find $x + y$. _____

Geometry                                    Student Technology Guide

NAME _____ CLASS _____ DATE _____

# Student Technology Guide
## 5.5 Special Triangles and Areas of Regular Polygons

In Lesson 5.4, you learned how to apply the Pythagorean Theorem to find side lengths in right triangles. Then in Lesson 5.5, you linked this skill to the problem of finding areas of regular polygons. This worksheet will help you use a calculator to carry out complicated calculations to solve an area problem.

**Example:** A regular hexagon is inscribed in a circle whose radius is 4. To the nearest tenth of a square unit, find the hexagon's area. The vertices of the hexagon and point O determine 6 congruent equilateral triangles and 12 congruent isosceles triangles. Thus, $AF = 4$ and $AX = 2$. As a result:

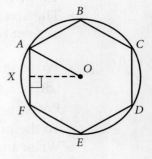

$$\text{area of hexagon } ABCDEF = 12 \times \text{area of } \triangle OAX$$
$$= 12(\tfrac{1}{2} \times 2 \times \sqrt{4^2 - 2^2})$$
$$= 12\sqrt{4^2 - 2^2}$$

Press 12 [2nd] [$x^2$] 4 [$x^2$] [ - ] 2 [$x^2$] [ ) ] [ENTER].
To the nearest tenth, the area is about 41.6 square units.

**A regular hexagon is inscribed in a circle with the given radius. To the nearest tenth of a square unit, find the area of the hexagon.**

1. $r = 10$ _____   2. $r = 24$ _____   3. $r = 36$ _____

When you study the diagram above of the regular hexagon inscribed in the circle, you may be able to see how to derive a formula for the area of the hexagon inscribed in a circle with a radius of $r$.

4. Write a formula for the area of a regular hexagon inscribed in a circle with a radius of $r$. _____

**Using the formula you found in Exercise 4, find the area of a regular hexagon inscribed in a circle with the given radius.**

5. $r = 10$ _____   6. $r = 24$ _____   7. $r = 36$ _____

8. a. A square is inscribed in a circle with a radius of 50. Write an expression for the area of the square. _____
   b. Use a calculator to approximate the area of the square to the nearest tenth of a square unit. _____

9. Write a formula for the area of a square inscribed in a circle with a radius of $r$. Use the formula to solve the problem in Exercise 8. _____

32    Student Technology Guide    Geometry

NAME _____ CLASS _____ DATE _____

# Student Technology Guide
## 5.6 The Distance Formula and the Method of Quadrature

In Lesson 5.6, you studied a method that can be used to approximate the area of a curved region. At right is a quarter of a circle on a coordinate plane, with 8 rectangles superimposed on it.

The width of each rectangle is $\frac{1}{8}$ of a unit.

The height of each rectangle is found by applying the Pythagorean Theorem to right triangles with a hypotenuse of 1. For example, the height of the fifth rectangle from the left is given by the equation below.

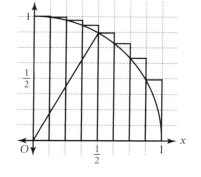

$$\text{height of fifth rectangle} = \sqrt{1^2 - \left(\frac{5-1}{8}\right)^2}$$

Then the area $A$ of the quarter circle is approximated by the sum of the areas of the 8 rectangles.

$$A \approx \frac{1}{8}\sqrt{1^2 - \left(\frac{1-1}{8}\right)^2} + \frac{1}{8}\sqrt{1^2 - \left(\frac{2-1}{8}\right)^2} + \cdots + \frac{1}{8}\sqrt{1^2 - \left(\frac{7-1}{8}\right)^2} + \frac{1}{8}\sqrt{1^2 - \left(\frac{8-1}{8}\right)^2}$$

Since the area of a circle with radius 1 equals $\pi$, you can find an approximation of $\pi$ by multiplying the approximate area of the quarter circle by 4. A spreadsheet is an ideal tool to use to approximate $\pi$.

**For the exercises below, use a spreadsheet.**

1. a. In a new spreadsheet, type column headings as shown at right.
   b. In cell A2, enter 1. In cell A3, enter =A2+1.
   c. In cell B2, enter the formula below.
      =SQRT(1-((A2-1)/8)^2)/8
   d. Fill columns A and B down to row 9 by selecting **Edit Fill Down**.
   e. In cell C2, enter = 4*SUM(B2:B9).

2. Explain why the entry 3.3398 is greater than $\pi$. By roughly how much is this entry greater than $\pi$?

_____

3. Explain how to modify the spreadsheet so that the total gives an approximation of $\pi$ by based on sum of the areas of 100 rectangles. Then give the approximation.

_____

_____

Geometry                                              Student Technology Guide    33

NAME _____ CLASS _____ DATE _____

# Student Technology Guide
## 5.7 Proofs Using Coordinate Geometry, page 1

Suppose that you are given quadrilateral ABCD with vertices $A(-3, -2)$, $B(-2, 3)$, $C(5, 5)$, and $D(4, 0)$. Your job is to analyze quadrilateral ABCD. Because the quadrilateral is defined by the coordinates of its vertices, you decide to use coordinate geometry.

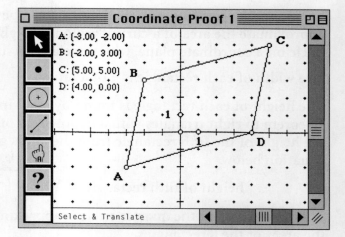

Once you establish a coordinate grid and sketch the quadrilateral, you can use measurement features to justify various conclusions.

Recall that you can sketch quadrilateral ABCD by
- using the point tool to place four points at the correct locations and then sketching the sides of ABCD,
- using the segment tool and placing endpoints of line segments at the specified locations, or
- selecting [Graph] [Plot Points...], entering the coordinates of A, B, C, and D, and then pressing [ADD].

**In Exercises 1–6, refer to the sketch above.**

1. Use a coordinate grid on your geometry graphics software to sketch quadrilateral ABCD.

The sketch above suggests that quadrilateral ABCD is a parallelogram. In Exercises 2–5:
a. Explain how you would use each approach to show that ABCD is a parallelogram.
b. Then use measurement features to carry out your strategy.

2. slopes of opposite sides _____

_____

3. lengths of opposite sides _____

_____

4. supplementary angles _____

_____

5. diagonals _____

_____

34      Student Technology Guide                                    Geometry

NAME _____ CLASS _____ DATE _____

# Student Technology Guide
## 5.7 Proofs Using Coordinate Geometry, page 2

The sketch at the right shows the same quadrilateral as in Exercises 1–5. However, the midpoints of the four sides have been constructed and labeled *W*, *X*, *Y*, and *Z*. These points determine quadrilateral *WXYZ*.

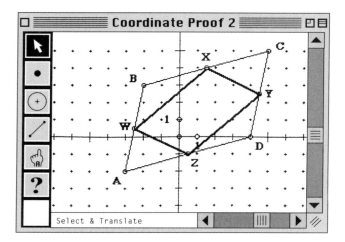

6. Outline a strategy, perhaps one of the strategies from Exercises 2–5, that you can use to classify quadrilateral *WXYZ*. Use measurement features of the software to carry out your strategy.

_____

_____

_____

The equations at right define three lines. Using geometry graphics software, you can graph these lines and analyze the geometric figure that results. To show an equation for a line, select the line, and then select  Measure   Equation . Adjust the line until the slope and *y*-intercept agree with those given, or construct a line through two points plotted by using  Graph   Plot Points .

$$\begin{cases} \overleftrightarrow{AB} : y = 0.57x + 1.86 \\ \overleftrightarrow{CD} : y = -2.00x - 7.00 \\ \overleftrightarrow{EF} : y = 1.60x + 0.80 \end{cases}$$

**For Exercises 7 and 8, refer to the sketch at right.**

7. Graph $\overleftrightarrow{AB}$, $\overleftrightarrow{CD}$, and $\overleftrightarrow{EF}$. Display the equations of the lines. Locate and mark points *X*, *Y*, and *Z*, where the lines intersect.

8. Analyze △*XYZ*.

_____

_____

_____

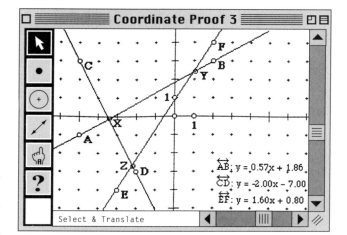

Geometry                                               Student Technology Guide      35

# Student Technology Guide
## 5.8 Geometric Probability

To find the probability that a point chosen at random in the rectangle at right will be inside the circle, you must find the ratio of the area of the circle to the area of the rectangle. Using geometry graphics software, you can easily sketch the circle and rectangle, calculate the area of each figure, and compute the ratio of the areas.

Also, the software enables you to enlarge or shrink the circle inside the rectangle. What effect does this have on the probability that a randomly chosen point will be inside the circle?

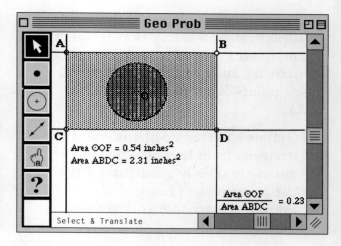

**Sketch each diagram below. Enlarge the circle to find the maximum probability that a point chosen at random will be inside the circle. (Hint: What is the largest radius the circle can have without breaking through the sides of the rectangle?)**

1.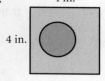
   4 in. × 4 in.

2.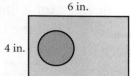
   6 in. × 4 in.

3.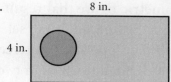
   8 in. × 4 in.

4. Suppose that a rectangle has a width of 4 and length of $a$, where $a \geq 4$. What happens to the maximum probability as $a$ increases? _____

5. Using the software, construct an equilateral triangle with sides of 5 inches. Place a circle inside it. Calculate the areas of the circle and triangle. Then calculate the ratio of their areas. Find the maximum probability that a point chosen at random within the triangle is also inside the circle.

6. Make a conjecture about the circle that maximizes the probability that a point chosen at random inside a regular polygon will also be inside the circle. Use the software to illustrate your conjecture for a regular pentagon.

# Student Technology Guide
## 6.3 Prisms, page 1

A cube is a square that is *extruded* in space so that the result is a square front and square back along with squares on the left, right, top, and bottom.

You can model a cube by starting with a square and translating it up and to the right. This imitates extrusion. Complete the drawing by joining the square and its image with line segments.

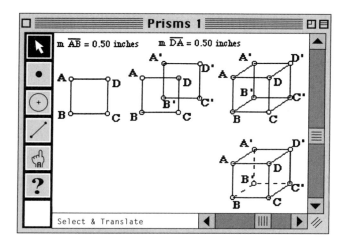

**In the exercises that follow, use geometry graphics software.**

1. a. Sketch square *ABCD* with two horizontal sides and two vertical sides. (Display the lengths of two adjacent sides to ensure you have drawn a square.)
   b. Select the square and choose the translation feature. Select a short length and an angle such as 35°. The image will be a square congruent to $A'B'C'D'$.
   c. Join $A$ and $A'$, $B$ and $B'$, $C$ and $C'$, and $D$ and $D'$. Use dashed lines for nonvisible edges. (Hide the point labels if desired.)

2. How do you know that square $ABCD \cong$ square $A'B'C'D'$ and that $\overline{AA'} \cong \overline{BB'} \cong \overline{CC'} \cong \overline{DD'}$?

_____

_____

3. Sketch rectangle *ABCD*. Use it and a translation of it to sketch the box at right.

4. How can you use the sketch from Exercise 3 to create the prisms shown here?

_____

_____

_____

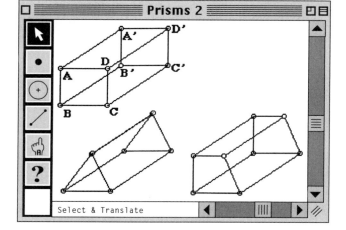

Geometry        Student Technology Guide

NAME _____ CLASS _____ DATE _____

# Student Technology Guide
## 6.3 Prisms, page 2

Once you sketch a box in your software, you can simulate rotating and flipping it by dragging a vertex around the screen. You can almost imagine it moving dynamically in space. In the sketch at right, the box modeled by the sketch on the left was moved around to produce the box modeled on the right. What actions really took place to get the box on the left to match the box to the right?

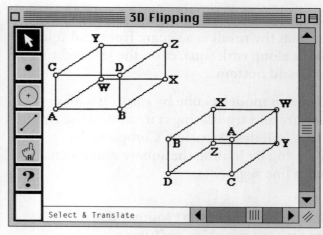

**Use geometry graphics software as needed.**

5. a. Using paper, cardboard, or some other material, make a box. Label the vertices as shown.
   b. Place your box on the desk in front of you so that the face labeled *ABDC* faces you, face *AWXB* lies on the desk, and face *DBXZ* is to your right.

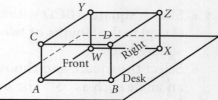

6. Flip or rotate the box on your desk to replicate the right side of the sketch above. What actions or transformations did you use to orient the box to make it match the display?

_____
_____
_____

7. Using the perpendicular line feature of the software, sketch rectangle *ABCD*. Then hide the lines used. Translate the rectangle to form the back of the box. Join front to back with line segments. Label the vertices to match the labels of the box shown on the left side of the sketch.

8. Drag a vertex of the box across the screen to show the box in a different orientation, keeping the front rectangular. Describe your actions with the box that correspond to the dragging on the screen.

_____
_____
_____
_____

38   Student Technology Guide                                    Geometry

NAME _____ CLASS _____ DATE _____

# Student Technology Guide
## 6.6 Perspective Drawing, page 1

Using either pencil and paper or geometry graphics software, you can represent an object in one-point perspective. However, there is an advantage to using software—you can drag the vanishing point $P$ to locations such as $P'$ and view the object from a different viewing point. Notice that both sketches in the display at right show the same box in perspective, but there are differences in the views.

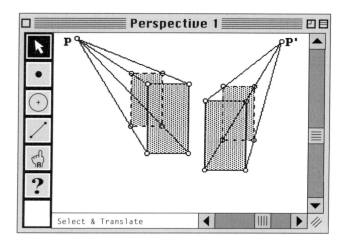

**Use geometry graphics software in the exercises below.**

1. a. Using the software, sketch a rectangle with two horizontal sides and two vertical sides.
   b. Place and label point $P$ outside the rectangle. Sketch line segments joining $P$ to each of the vertices of the rectangle.
   c. Place a point on one of the line segments drawn. Draw one vertical and one horizontal line segment with this point as one endpoint and the other endpoints on two of the line segments drawn in part **b**. Draw the other two line segments that complete the back face of the box.

2. The sketch drawn in Exercise 1 is shown on the left side of the sketch window. A copy was placed on the right side. Describe the locations of a viewer looking at the sketch on the left and of a viewer looking at the sketch on the right.

   _____

   _____

3. Suppose that you want to show the box with the vanishing point in some location other than $P$ or $P'$. Where would you drag $P$ so that the viewer is looking straight at the box, unable to see any face of the box other than the front? Modify your sketch to illustrate. _____

   _____

4. Experiment with other locations for the vanishing point. Describe some of the resulting views. _____

   _____

   _____

Geometry                                    Student Technology Guide    39

# Student Technology Guide
## 6.6 Perspective Drawing, page 2

The sketch at right shows a building in two-point perspective. The building has two windows and one door, each shown in black.

**Use geometry graphics software and two-point perspective in the exercises below.**

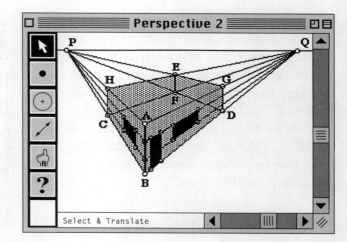

5. Without being given any instructions, replicate the diagram of the building with its windows and door in two-point perspective. Your finished sketch need not have the exact dimensions or measures shown here. (Hint: Start by drawing a horizontal line $\overleftrightarrow{PQ}$ and a vertical line segment $\overline{AB}$ below $\overleftrightarrow{PQ}$.)

6. Suppose that you want faces *HCBA* and *ABDG* to appear congruent. Explain how to modify the sketch you drew in Exercise 5 to accomplish this. Then perform the modification.

_____

_____

_____

The diagram below shows the front view of a house whose roof is shaped like an isosceles trapezoid.

7. Using the software and either one-point or two-point perspective, make a new sketch of the house from a different viewpoint. Summarize your work and strategies.

_____

_____

_____

_____

_____

_____

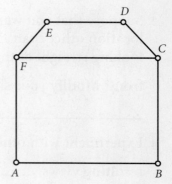

**40**  Student Technology Guide  Geometry

NAME _____ CLASS _____ DATE _____

# Student Technology Guide
## 7.3 Surface Area and Volume of Pyramids, page 1

Suppose that you and some classmates have been asked to participate in designing a new park monument–a tall pyramid sitting on top of a solid rectangular base. The following relationships exist among the dimensions:

$$x' = \tfrac{1}{2}x \qquad h' = 10h \qquad h = \tfrac{1}{4}x$$

Your job is to use the diagram at the right to carry out the calculations needed for the construction of the monument. Recall the formulas below.

Volume of a prism = height × area of base

Volume of a pyramid = $\tfrac{1}{3}$ × height × area of base

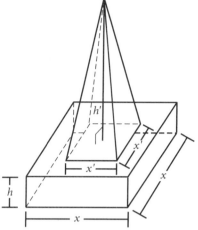

**Use a calculator as needed. Round your answers to the nearest hundredth.**

1. a. Show that $V_b$, the volume of the base of the monument, can be written as $\tfrac{1}{4}x^3$. _____

   b. Show that $V_p$, the volume of the pyramidal top of the monument, can be written as $\tfrac{10}{48}x^3$. _____

   c. Show that $V$, the volume of the entire monument, can be written as $\tfrac{22}{48}x^3$. _____

To evaluate $\tfrac{22}{48}x^3$ on a calculator for a value of $x$, such as 5, press 22 [÷] 48 [×] 5 [^] 3 [ENTER].

2. Find the volume of the monument for each value of $x$ below.
   a. 10 feet        b. 12 feet        c. 18 feet        d. 20 feet

   _____   _____   _____   _____

3. The monument is to be made of concrete. If $x$ is given in feet, then $\tfrac{22}{48}x^3$ gives the volume in cubic feet. Modify your expression so that the volume is in cubic yards. (cubes with an edge length of one yard) _____

   Find the volume in cubic yards if $x$ is 15.5 feet. _____

The visible surfaces of the monument will be covered with copper sheathing. Using the relationships among height and length, you can express the monument's visible surface area as $\left(\tfrac{7 + \sqrt{101}}{4}\right)x^2$. Recall that [2nd] [$x^2$] accesses the square root function.

4. Find the visible surface area of the monument for each value of $x$.
   a. 10 feet        b. 12 feet        c. 18 feet        d. 20 feet

   _____   _____   _____   _____

Geometry                                                Student Technology Guide    41

NAME _____ CLASS _____ DATE _____

# Student Technology Guide
## 7.3 Surface Area and Volume of Pyramids, page 2

Some community members have proposed a very different design for the park monument. It is pictured at the right, but not to scale. Relative dimensions of the top slab are the same as those of the bottom slab. Each vertical column has a square base with a side length of $0.1x$ units and height of $1.5x$ units.

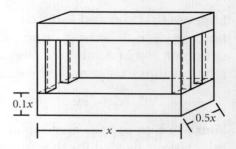

**Use a calculator as needed.**

5. Explain how you know that the following expression represents the total volume of the monument: $2[(0.1x)(x)(0.5x)] + 4[(0.1x)^2(1.5x)] = 0.16x^3$

_____

_____

_____

Use a graphics calculator's features below to graph $y = 0.16x^3$.

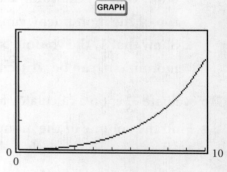

6. Find the volume of the monument for each value of $x$ below. (Press TRACE, then use the arrow keys until the given value of $x$ is displayed on the screen.)

   a. 6 feet         b. 7 feet         c. 8 feet         d. 9 feet

   _____     _____     _____     _____

If you know the volume of the monument, you can approximate the value of $x$ on the graph for a given value of $y$.

7. Use a graphics calculator to approximate $x$ for each given value of $y$. Round your answers to the nearest tenth.

   a. 110 cubic feet     b. 120 cubic feet     c. 130 cubic feet     d. 140 cubic feet

   _____         _____         _____         _____

# Student Technology Guide
## 7.4 Surface Area and Volume of Cylinders

You can make a pipe by constructing an *annulus*, which is a pair of distinct coplanar circles with the same center, and then *extruding* the annulus straight up into three-dimensional space.

The surface area $S$ of a pipe depends on its length, $h$. It also depends on both the inside radius, $s$, and the outside radius, $r$, of the annulus. The volume $V$ of the material used to make the pipe also depends on the thickness, $t$, of the pipe.

Using a spreadsheet, you can efficiently calculate the surface area and volume of pipes of varying thickness.

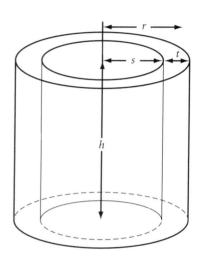

**Refer to the spreadsheet below. It has data for a pipe with an outside radius of 3 inches and a length of 10 feet (120 inches). The thickness of the pipe varies.**

1. Set up columns A, B, C, and D in a spreadsheet like the one shown.

2. a. Write a formula in terms of $r$, $t$, and $h$ for the surface area of the pipe and the volume of the material used to make the pipe.

   $S = $ _____    $V = $ _____

   b. Write spreadsheet formulas for $S$ and $V$, and enter them into cells E2 and F2. Then **FILL** **DOWN** columns E and F.

   E2: _____    F2: _____

3. Both $S$ and $V$ are functions of the thickness $t$. Describe how $S$ and $V$ change as $t$ decreases.

   _____

An *open can* is a pipe whose bottom is plugged by a thin cylinder with a radius of $s$ and height of $h'$.

4. Write formulas for the surface area and volume of an open can.

   $S = $ _____

   $V = $ _____

5. Use a spreadsheet to explore how $S$ and $V$ change given the information above and $h' = t$.

|   | A | B | C | D | E | F |
|---|---|---|---|---|---|---|
| 1 | r | t | s | h | S | V |
| 2 | 3 | 0.20 | 2.80 | 120.00 | | |
| 3 | 3 | 0.19 | 2.81 | 120.00 | | |
| 4 | 3 | 0.18 | 2.82 | 120.00 | | |
| 5 | 3 | 0.17 | 2.83 | 120.00 | | |
| 6 | 3 | 0.16 | 2.84 | 120.00 | | |
| 7 | 3 | 0.15 | 2.85 | 120.00 | | |
| 8 | 3 | 0.14 | 2.86 | 120.00 | | |
| 9 | 3 | 0.13 | 2.87 | 120.00 | | |
| 10 | 3 | 0.12 | 2.88 | 120.00 | | |
| 11 | 3 | 0.11 | 2.89 | 120.00 | | |
| 12 | 3 | 0.10 | 2.90 | 120.00 | | |

_____

NAME _____ CLASS _____ DATE _____

# Student Technology Guide
## 7.5 Surface Area and Volume of Cones

The following example shows how to use a calculator to find the volume of a cone:

**Example:** A cone has a height of 18 inches and a base radius of 8 inches. What is the volume of the cone?

Evaluate $V = \frac{1}{3}\pi r^2 h$, or $V = \frac{\pi}{3}r^2 h$.

Press [2nd] [^] [÷] 3 [×] 8 [$x^2$] [×] 18 [ENTER].

The volume of the cone is about 1206.37 cubic inches.

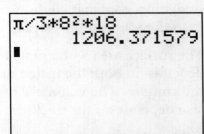

**To the nearest tenth of a cubic unit, find the volume of each cone described below.**

1. $r = 10$ and $h = 20$ _____
2. $r = 10.5$ and $h = 200$ _____
3. $r = 24$ inches and $h = 24$ inches _____
4. $r = 10$ feet and $h = 1$ foot _____

A *frustum* of a cone is the section that is left when the top of the cone is cut off by a plane parallel to the plane containing the cone's base. In the figure at right, the shaded frustum results when a cone whose height is $h'$ units is removed. The radius and height of the removed cone are proportional to the radius and height of the full cone.

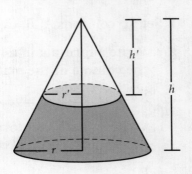

**Example:** A cone has a height of 18 inches and a base radius of 8 inches. The top quarter of the cone is removed. Find the volume of the frustum.

Evaluate $V = \frac{1}{3}\pi r^2 h - \frac{1}{3}\pi (r')^2 h'$, given $r' = \frac{1}{4}r$ and $h' = \frac{1}{4}h$.

- Evaluate $V = \frac{1}{3}\pi (8^2)(18) - \frac{1}{3}\pi\left(\frac{8}{4}\right)^2\left(\frac{18}{4}\right)$, or $\frac{\pi}{3}\left[18(8^2) - \left(\frac{8}{4}\right)^2\left(\frac{18}{4}\right)\right]$.

Press [2nd] [^] [÷] 3 [×] [(] 18 [×] 8 [$x^2$] [−] [(] 8 [÷] 4 [)] [$x^2$] [×] [(] 18 [÷] 4 [)] [)] [ENTER].

The volume of the frustum is about 1187.5 cubic inches.

**A cone has a height of 18 inches and a base radius of 8 inches. Approximate the volume of each frustum if the following parts of the cone are removed:**

5. top half

6. top tenth

7. top two-thirds

8. top 5%

9. top 10%

10. top 90%

44    Student Technology Guide    Geometry

NAME _____ CLASS _____ DATE _____

# Student Technology Guide
## 7.6 Surface Area and Volume of Spheres

To find the surface area $S$ and volume $V$ of a sphere with radius $r$, apply the formulas $S = 4\pi r^2$ and $V = \frac{4}{3}\pi r^3$. You can use them to find the surface area and volume of a spherical lune such as the one shown at right. The lune is a two-dimensional surface determined by the angle marked $\theta$. To find the surface area of a lune and the volume of its associated spherical wedge, see the calculator example below.

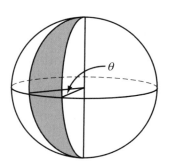

**Example:** Find the surface area of the lune and the volume of a wedge formed by a 60° angle in a sphere whose radius is 12 inches.

Since 60° is $\frac{1}{6}$ of a complete revolution, find one-sixth of the surface area and one-sixth of the volume of the sphere.

- Surface area: $\frac{4\pi r^2}{6} = \frac{2\pi r^2}{3}$; Volume: $\frac{\frac{4}{3}\pi r^3}{6} = \frac{2\pi r^3}{9}$

- Press 2 [2nd] [^] [×] 12 [$x^2$] [÷] 3 [ENTER]. Press [2nd] [ENTER] to edit the expression. Use [◄] to replace the exponent 2, and press [^] [2nd] [DEL] 3. Use [►] to replace the denominator, 3. Press 9 [ENTER].

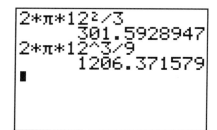

- The lune covers about 301.6 in.² The wedge holds about 1206.4 in.³

The figures and formulas at right are for a spherical cap and a spherical zone.

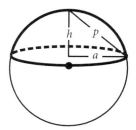

$S = \pi p^2$
$V = \frac{h\pi p^2}{6}$

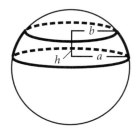

$S = 2r\pi h$
$V = \frac{3a^2 + 3b^2 + h^2}{6}\pi h$

A sphere has a radius of 12 inches. Find the surface area and volume of each shape to the nearest tenth, according to the given information.

1. $\theta = 45°$

2. $h = 3, a = 5, p = \sqrt{34}$

3. $h = 3, a = 5, b = 2$

4. $h = \sqrt{10}, a = \sqrt{20}, b = \sqrt{8}$

5. $\theta = 120°$

6. $h = 12, a = 12, p = 12\sqrt{2}$

Geometry                                Student Technology Guide   **45**

# Student Technology Guide
## 8.1 Dilations and Scale Factors, page 1

One day Debra sketched △ABC, located and labeled point O outside of △ABC, chose O as a center of dilation, and then dilated △ABC. Without thinking, she chose a negative number, −1.5, as the scale factor. Shown here is the image △A'B'C' that she got.

It appears that △A'B'C' is related to △ABC in another way. Whether the scale factor is positive or negative is important.

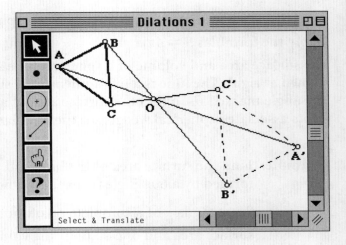

**Use geometry graphics software as directed.**

1. Sketch △ABC with heavy lines and point O as shown. Using O as center of dilation and scale factor −1.5, construct the image of △ABC with dashed lines. Show its vertex labels.

2. Based only on observations, describe how A'B'C' is obtained from △ABC.

3. Modify △ABC or drag point O to make different dynamic drawings. Do your observations from Exercise 2 change or remain the same? Explain your response.

4. Use experimentation, reasoning, and the software to complete the summary table below.

| Scale factor, s | Description of transformation |
|---|---|
| $s > 1$ | |
| $s = 1$ | |
| $0 < s < 1$ | |
| $s = 0$ | |
| $-1 < s < 0$ | |
| $s = -1$ | |
| $s < -1$ | |

46    **Student Technology Guide**    Geometry

NAME _____ CLASS _____ DATE _____

# Student Technology Guide
## 8.1 Dilations and Scale Factors, page 2

Debra tries another experiment. She asked the following question.
  *What happens if a dilation is performed successively?*
In this experiment, you will use a coordinate approach to answer the question.

**Use geometry graphics software in the exercises below.**

5. a. Open a new sketch. In the main menu, select Graph Show Grid.
   b. Using the point tool, place a point whose coordinates are $A(2, 1)$. Show the point label. Relabel if necessary. Select the point and Measure Coordinates from the main menu. Move the display of the coordinates aside.

6. Let $s = 0.75$. Select O as the center of dilation. Select Transform Mark Center "O".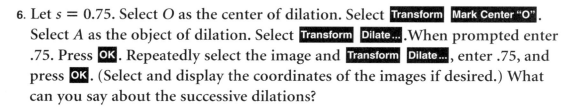
Select A as the object of dilation. Select Transform Dilate…. When prompted enter .75. Press OK. Repeatedly select the image and Transform Dilate…, enter .75, and press OK. (Select and display the coordinates of the images if desired.) What can you say about the successive dilations?

_____

_____

7. Repeat Exercise 6 for $s = 1.25$. How is the result similar to and different from the result from Exercise 6?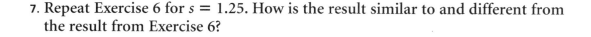

_____

_____

8. Debra wants to experiment with dilations to make point A bounce back and forth between Quadrant I and Quadrant III. She wants the images to get farther apart as the dilations proceed. What value of $s$ might she use? Illustrate your answer with a software sketch.

_____

_____

Geometry                                        Student Technology Guide    47

NAME _____ CLASS _____ DATE _____

# Student Technology Guide
## 8.2 Similar Polygons

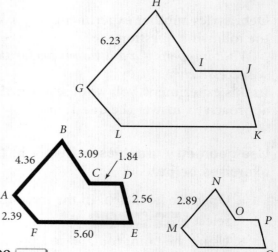

In the diagram at right are polygons GHIJKL and MNOPQR, each similar to polygon ABCDEF but with different scale factors. Enough information is given to find the unknown lengths of all the sides of polygons GHIJKL and MNOPQR. Five equations are needed to find the lengths in GHIJKL. Is there a way to simplify the process by using a calculator? The example below indicates that the answer is yes.

The factor relating GHIJKL to ABCDEF is $\frac{HG}{BA}$ or $\frac{6.23}{4.36}$. Multiply each length in ABCDEF by $\frac{HG}{BA}$ to find the corresponding length in GHIJKL.

**Example:**
- To find HI, enter 6.23 [÷] 4.36 [×] 3.09 [ENTER].
- To find IJ, edit the expression for HI. Press [2nd] [ENTER], replace 3.09 with 1.84, and then press [ENTER].
- To find the lengths of each of the other sides of GHIJKL, proceed as in the previous step. Replace 1.84 with 2.56, 2.56 with 5.60, and finally 5.60 with 2.39.
To the nearest hundredth of a unit, HI ≈ 4.42, IJ ≈ 2.63, JK ≈ 3.66, KL ≈ 8.00, and LG ≈ 3.42.

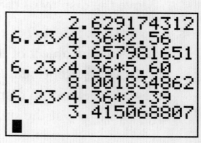

In the exercises that follow, use a calculator to find the unknown lengths. Round your answers to the nearest hundredth.

1. a. Using the ratio $\frac{2.89}{4.36}$, find MN above. _____
   b. Using your expression for MN and the example above as a model, find NO, OP, PQ, and QR. _____

2. Find XY, YZ, and ZW.    3. Find PQ, QR, RS, and ST.

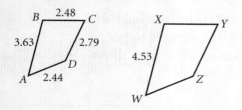

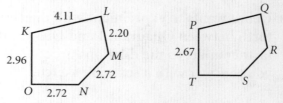

4. In △ABC above, AB = 10, BC = 20, and AC = 15. If △ABC ~ △XYZ and XY = 54, find YZ and XZ. _____

48    Student Technology Guide    Geometry

NAME _____ CLASS _____ DATE _____

# Student Technology Guide
## 8.4 The Side-Splitting Theorem

Suppose that someone asserts the following:

  A line parallel to one side of a triangle divides the triangle into similar triangles.

You can sketch △ABC as shown, locate and label an arbitrary point, D, on one side (such as $\overline{AB}$), sketch a line through point D parallel to $\overline{BC}$, and the locate and label point E where the line intersects $\overline{AC}$. This diagram can be used to confirm or deny the claim.

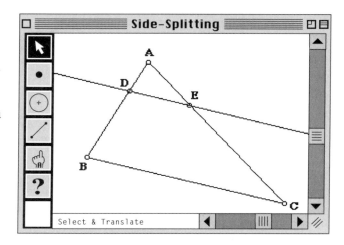

**Use geometry graphics software as needed.**

1. In your software, make a sketch according to the instructions above.

2. To verify the similarity claim above, what data would you gather from the diagram? Find and display the data. What similarity postulate or theorem did you use?

   _____

   _____

3. Experiment to find out whether the placement of the line parallel to $\overline{BC}$ affects the truth of the claim. If your investigations indicate that the claim is not stated with sufficient precision, rewrite the claim so that it is.

   _____

   _____

Consider the assertion below.
  *If △ABC is dilated with a vertex as the center of the dilation and positive scale factor, then the image and preimage form a pair of nesting triangles that are similar.*

4. a. Use geometry graphics software to represent the hypothesis of the assertion.
   b. To verify the similarity claim above, what data would you gather about the diagram? Find and display the data. What similarity postulate or theorem did you use?

   _____

   _____

Geometry                                              Student Technology Guide   **49**

NAME _____ CLASS _____ DATE _____

# Student Technology Guide
## 8.5 Indirect Measurement and Additional Similarity Theorems, page 1

A civil engineer marks point B, measures a distance due east, and stops at point C. The distance is 94 meters. Then she moves due east to point E. The distance between C and E is 39 meters. From point E, she moves due south 53 meters, arriving at point F. She does this in order to find AB, the distance across a river that runs perpendicular to $\overline{AB}$.

With geometry graphics software, you can model the data in a scale drawing and use the scale to find AB.

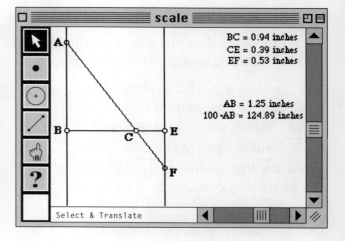

**In Exercises 1–4, 1 inch corresponds to 100 meters.**

1. a. In a new sketch, draw a horizontal line segment. Display and label its endpoints, B and C. Select B and C, and find the distance between them. Adjust C so that BC is 0.94 inches.
   b. Starting at point C, draw a horizontal line segment to the right. Display and label the right endpoint E. Adjust E so that the distance between C and E is 0.39 inches.
   c. Construct a line perpendicular to $\overline{CE}$ at E. Place a point below $\overline{CE}$ on this line, and label it F. Adjust F so that EF is 0.53 inches.
   d. Select B and $\overline{BC}$. Construct a line perpendicular to $\overline{BC}$ at B.
   e. Draw a line segment starting at F and passing through C. Extend the line segment until it meets the line from part **d**. Label the endpoint A.

2. The actual distance between A and B is 100 times the distance between A and B in the sketch. Find AB in the sketch. Then use the calculation feature to find 100 × AB.

3. Given $\triangle ABC \sim \triangle FEC$, explain how two similar objects were used to find AB.

_____

4. In your sketch, delete AB and 100 × AB. Modify the sketch so that AB = 1.20, BC = 0.80, and CE = 0.60. Assuming the same scale, what are the actual distances represented by AC and CF? _____

5. Draw a new sketch. Points A, B, C, D, and E lie in the same plane. The distance between A and C is 164 units. Point D is directly above A, and AD measures 78 units. Point E is 90 units directly below C. $\overline{AC}$ and $\overline{DE}$ intersect at point B. Choose your own scale for the drawing, and find the actual distances between D and B, B and E, and B and C.

_____

50   Student Technology Guide   Geometry

NAME _____ CLASS _____ DATE _____

# Student Technology Guide
## 8.5 Indirect Measurement and Additional Similarity Theorems, page 2

A kitchen designer often uses software to make a scale drawing that represents a renovation plan.

A simple sketch of a kitchen plan is shown at right.

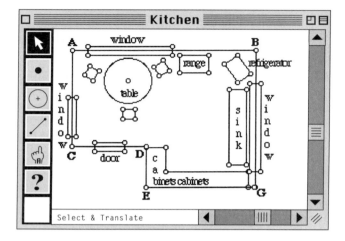

**Use geometry graphics software.**

6. A homeowner comes to you with the measurements shown in the box at right and asks for a scale drawing of the room represented in the sketch. Your job is to make a scale drawing.
   a. Choose a reasonable scale for the model. Write the scale.

   1 inch represents _____ feet.
   b. Sketch *ABGEDC*. Adjust the vertices so that all corners are right angles.
   c. Adjust the lengths so that all sides of the room are to scale.

$AB$ = 15 feet 6 inches
$BG$ = 12 feet 0 inches
$GE$ = 8 feet 3 inches
$ED$ = 4 feet 6 inches
$DC$ = 7 feet 3 inches
$CA$ = 7 feet 6 inches

**For each object listed below, write its dimensions according to your scale from part a of Exercise 6. Then add a representation of the object to the scale drawing.**

7. along $\overline{BG}$, a window 7 feet long with negligible depth _____

8. along $\overline{AC}$, a window 3 feet 9 inches long with negligible depth _____

9. along $\overline{AB}$, a window 6 feet long with negligible depth _____

10. along $\overline{DE}$ and $\overline{EG}$, cabinets that are 24 inches deep _____

11. along $\overline{BG}$, a sink/counter 6 feet long, 2 feet 6 inches deep _____

12. along $\overline{CD}$, a door 3 feet wide _____

13. The homeowner wants to cover the entire floor with tile.

   a. Find the area of *ABGEDC*. _____

   b. How many square feet does 1 square inch represent? _____

   c. Use **Measure Calculate...** to find how many square feet of tile are needed. _____

Geometry      Student Technology Guide

NAME _____ CLASS _____ DATE _____

# Student Technology Guide
## 8.6 Area and Volume Ratios

From your study of similar figures, you know the following about similar figures with similarity ratio $s$:   area of shape II $= s^2$(area of shape I)
This relationship comes into play when someone analyzes a quantity such as rate of flow through a pipe or series of pipes. One such series is shown below.

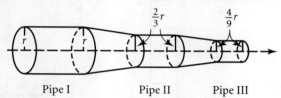

Pipe I     Pipe II     Pipe III

If a circular pipe has an inside radius of $r$, then the *rate of flow F* of a liquid (volume per unit time) is given by the formula at right. In the formula, $k$ depends on the fluid.

$$F = \left(\frac{k}{2}\right) \times \text{cross-sectional area} \times r^2$$

To compare two rates of flow, $F_1$ and $F_2$, through two pipes, consider $\frac{F_1}{F_2}$.

Suppose that pipes $P_1$ and $P_2$ have circular cross sections and that the radius of a circular cross section of $P_2$ is $s$ times the radius $r$ of a cross section of $P_1$.

$$\frac{F_1}{F_2} = \frac{\left(\frac{k}{2}\right) \times \text{area of } P_1 \times (\text{radius of } P_1)^2}{\left(\frac{k}{2}\right) \times \text{area of } P_2 \times (\text{radius of } P_2)^2} \left(\frac{\text{area of } P_1}{\text{area of } P_2}\right)\left(\frac{\text{radius of } P_1}{\text{radius of } P_2}\right)^2 = \left(\frac{1}{s^2}\right)\left(\frac{r}{rs}\right)^2 = \frac{1}{s^4}$$

**Example:** Compare the flow rate through pipe $P_1$ to the flow rate through pipe $P_2$ if the radius of $P_2$ is 80% that of pipe $P_1$.
- The scale factor $s$ is 0.8 since 80% equals 0.8.
  Find $\frac{1}{0.8^4}$.
- Press 1 [÷] .8 [^] 4 [ENTER].
Pipe $P_1$'s flow rate is about 2.44 times the flow rate through pipe $P_2$.

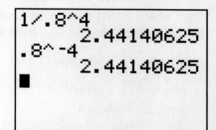

### Find each comparison of $F_1$ to $F_2$.

1. $P_1$:$r$ and $P_2$:$0.50r$ _____     2. $P_1$:$r$ and $P_2$:$2.00r$ _____

3. Pipe 1 has radius $r$ and flow rate $F_1$, pipe 2 has radius $\frac{2}{3}r$ and flow rate $F_2$, and pipe 3 has radius two-thirds that of pipe 2 and flow rate $F_3$. Find $F_1$:$F_2$, $F_2$:$F_3$, and $F_1$:$F_3$. _____

4. If $\frac{F_1}{F_2} = 16$, evaluate $\sqrt[4]{16}$ to find how radii are related. _____

52     Student Technology Guide     Geometry

NAME _____ CLASS _____ DATE _____

# Student Technology Guide
## 9.1 Chords and Arcs, page 1

Some people think that the circle is the only perfect planar figure. There is some truth to this. You may be surprised to find out that you can scatter *n* points around a circle and use them to construct a regular polygon with *n* sides. At the right are the beginning sketches of an equilateral triangle and a regular pentagon. Notice that the sketches also display related arc measures.

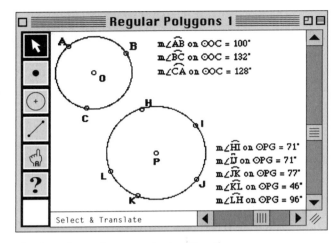

**For the exercises below, use geometry graphics software as directed.**

1. a. In a new sketch, draw two circles. On one circle, place three points. On the other circle, place five points. Display the point labels. (To place a point on the circle, select the circle and `Construct` `Point On Object`.)

   b. Find and display the measures of the three arcs on the first circle and the five arcs on the other circle. (To find and display the measure of an arc, select the two endpoints of the arc and select the circle. From the main menu, select `Measure` `Arc Angle`.)

2. a. Consider the circle with three points on it. Explain how to adjust the arc measures to make $\overline{AB} \cong \overline{BC} \cong \overline{CA}$. _____

   b. Carry out your thinking. Create and classify $\triangle ABC$. _____

3. a. Consider the circle with five points on it. Explain how to adjust the arc measures to make $\overline{HI} \cong \overline{IJ} \cong \overline{JK} \cong \overline{KL} \cong \overline{LH}$.
   _____

   b. Carry out your thinking. Create and classify pentagon *HIJKL*. _____
   _____

4. Evaluate the following strategy for constructing a regular polygon with *n* sides: ① Scatter *n* points on a circle. ② Find the measures of the *n* arcs determined. ③ Find the average of the *n* measures. ④ Adjust each arc so that its measure equals the average found. ⑤ Join the *n* points in order around the circle.
   _____
   _____
   _____

Geometry                                            Student Technology Guide

NAME _____ CLASS _____ DATE _____

# Student Technology Guide
## 9.1 Chords and Arcs, page 2

The distance that a point travels around a circle depends on the radius of the circle and the angle through which the point travels. The diagram below illustrates three points traveling around point O in a circular path through an angle $\theta$.

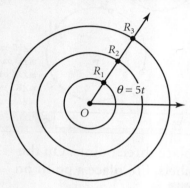

| | A | B | C | D | E | F | G | H |
|---|---|---|---|---|---|---|---|---|
| 1 | T | $\theta$ | R1 | R2 | R3 | D1 | D2 | D3 |
| 2 | 0 | 0 | 4 | 8 | 12 | 0.00 | 0.00 | 0.00 |
| 3 | 5 | 25 | 4 | 8 | 12 | 1.75 | 3.49 | 5.24 |
| 4 | 10 | 50 | 4 | 8 | 12 | 3.49 | 6.98 | 10.47 |
| 5 | 15 | 75 | 4 | 8 | 12 | 5.24 | 10.47 | 15.71 |
| 6 | 20 | 100 | 4 | 8 | 12 | 6.98 | 13.96 | 20.94 |
| 7 | 25 | 125 | 4 | 8 | 12 | 8.73 | 17.45 | 26.18 |
| 8 | 30 | 150 | 4 | 8 | 12 | 10.47 | 20.94 | 31.42 |
| 9 | 35 | 175 | 4 | 8 | 12 | 12.22 | 24.43 | 36.65 |
| 10 | 40 | 200 | 4 | 8 | 12 | 13.96 | 27.93 | 41.89 |
| 11 | 45 | 225 | 4 | 8 | 12 | 15.71 | 31.42 | 47.12 |
| 12 | 50 | 250 | 4 | 8 | 12 | 17.45 | 34.91 | 52.36 |
| 13 | 55 | 275 | 4 | 8 | 12 | 19.20 | 38.40 | 57.60 |
| 14 | 60 | 300 | 4 | 8 | 12 | 20.94 | 41.89 | 62.83 |
| 15 | 65 | 325 | 4 | 8 | 12 | 22.69 | 45.38 | 68.07 |
| 16 | 70 | 350 | 4 | 8 | 12 | 24.43 | 48.87 | 73.30 |
| 17 | 75 | 375 | 4 | 8 | 12 | 26.18 | 52.36 | 78.54 |

The spreadsheet at right records the following information:
- elapsed time $t$ in seconds
- angle of travel $\theta$ in degrees ($\theta = 5t$)
- radii: $R_1 = 4$, $R_2 = 8$, and $R_3 = 12$
- travel distances: $D_1$, $D_2$, and $D_3$

**For Exercises 5–7, refer to the spreadsheet above.**

5. a. What does the entry in cell B17 tell you? _____
   b. What do the entries in cells F17, G17, and H17 tell you about how $D_1$, $D_2$, and $D_3$ are related? _____

6. Find the ratios $R_2$ to $R_1$ and $R_3$ to $R_1$ and the ratios $D_2$ to $D_1$ and $D_3$ to $D_1$. What do your answers tell you about distance and radius?

   _____

7. The formulas entered into cells F2, G2, and H2 are shown at right. Create a spreadsheet to find $D_1$, $D_2$, and $D_3$ given $R_1 = 5$, $R_2 = 10$, and $R_3 = 15$. Is it true that if three points travel through an angle $\theta$ around concentric circles and radius $R_2$ is

   F2: =B2*3.14159*C2/180
   G2: =B2*3.14159*D2/180
   H2: =B2*3.14159*E2/180

   twice $R_1$ and radius $R_3$ is three times $R_1$, then $D_2 = 2D_1$ and $D_3 = 3D_1$? _____

8. Circles $C_1$ and $C_2$ have radii $R$ and $Rs$, where $s > 0$, and each point travels around each circle through an angle $\theta$. How do the distances traveled compare, and which point travels farther?

   _____

   _____

**54** Student Technology Guide  Geometry

NAME _____ CLASS _____ DATE _____

# Student Technology Guide
## 9.2 Tangents to Circles, page 1

When you draw a set of congruent central angles in a given circle, you have a fascinating exploration ahead. Continue it by completing the exercises below.

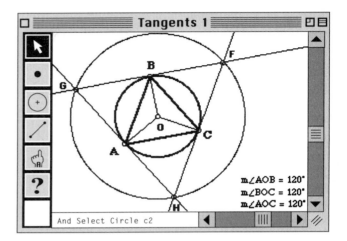

**Use geometry graphics software in the exercises below.**

1. a. Sketch a circle and place three distinct points, *A*, *B*, and *C*, on it. Find and display the measures of the central angles ∠*AOB*, ∠*BOC*, and ∠*AOC*. Move *B* and *C* until the measures of all three angles are equal.
   b. Using the fact that a tangent is perpendicular to a radius at the point where the radius intersects the circle, sketch tangents at *A*, *B*, and *C*. Construct and label the points where two tangents intersect. In this diagram, the points are labeled *F*, *G*, and *H*.
   c. Construct a circle with center *O* passing through *F*, *G*, and *H*. (To construct the circle, select *O*, then select one of the points, *F*, *G*, or *H*, then `Construct Circle By Center + Point`.)

2. Create and classify △*ABC*. Justify your response. _____

3. a. Make an educated guess about the relationship between △*ABC* and △*GHF*. _____
   b. Use measurement features of the software to confirm your guess from part a. Justify your work.

   _____

   _____

4. Suppose that you draw tangents to the larger circle at points *G*, *H*, and *F* and that you mark the points where these tangents intersect. What can you say about the relationship between △*ABC*, △*GHF*, and the triangle just determined? Justify your conjecture. Do you think that the pattern would continue indefinitely with the construction of more circles centered at *O* and more tangents to the circles?

   _____

   _____

Geometry                                                                                        Student Technology Guide

NAME _____ CLASS _____ DATE _____

# Student Technology Guide
## 9.2 Tangents to Circles, page 2

Just as you can place three points *A*, *B*, and *C* on a circle and explore the triangles that result, you can add a fourth point *D* to the diagram and explore the quadrilaterals that are formed.

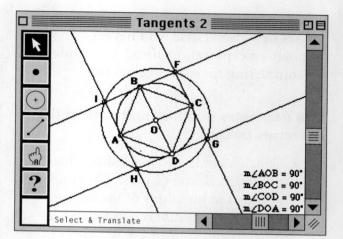

5. Modify the dynamic sketch that you made in Exercise 1 so that it resembles the sketch shown at the right.

6. Classify *ABCD*. _____

7. a. Make an educated guess about the relationship between quadrilaterals *ABCD* and *GHIF*.

   _____

   b. Use the software's measurement features to confirm your guess in part a. Justify your work.

   _____

   _____

8. Suppose that you draw tangents to the larger circle at points *G*, *H*, *I*, and *F* and that you mark the points where the tangents intersect. What can you say about the relationship between the quadrilateral just determined and quadrilaterals *ABCD* and *GHIF*? Justify your conjecture. Do you think that the pattern would continue indefinitely with the construction of more circles centered at *O* and more tangents to the circles?

   _____

   _____

In mathematical investigations, it is not unusual to explore a problem involving triangles, extend the problem to one involving quadrilaterals, and finally extend that exploration to polygons with *n* sides.

9. Suppose that you place *n* points on a given circle in such a way that they determine *n* congruent central angles. What cay you say about the sequence of *n*-gons (polygons with *n* sides) formed by successive constructions of intersecting tangents?

   _____

   _____

56   Student Technology Guide                                           Geometry

NAME _____ CLASS _____ DATE _____

# Student Technology Guide
## 9.3 Inscribed Angles and Arcs

Since geometry graphics software allows you to measure angles and arcs, you will find it helpful in studying problems that involve inscribed angles.

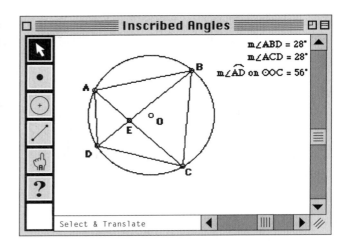

To find the measure of $\widehat{AD}$, select A, the circle, and then D. From the main menu, select  Measure   Arc Angle .

**Use geometry graphics software in the exercises below.**

1. Create a diagram something like the one at right. Find and display m∠ABD, m∠ACD, and m$\widehat{AD}$. Do your measurements confirm a theorem that you learned in Lesson 9.3? What theorem?

   _____

2. Display AB, DC, AE, DE, BE, and CE. Calculate the ratios AB:DC, AE:DE, and BE:CE. What can you conclude about △ABE and △DCE? Explain.

   _____

3. Devise and carry out a method different from the method in Exercise 2 to find out how △ABE and △DCE are related. What relationship did you discover?

   _____

4. Can you draw the same conclusion about △AED and △BEC as you did in Exercise 2? Explain. _____

   _____

5. Move points A, B, C, and D around the circle until point E coincides with point O, the center of the circle. What can you say about ∠ABC, ∠BCD, ∠CDA, and ∠DAB? What figure is formed by points A, B, C, and D?

   _____

6. Find and display m$\widehat{BCD}$ + m$\widehat{CDA}$ + m$\widehat{DAB}$ + m$\widehat{ABC}$. How does this calculation help confirm that the sum of the measures of the interior angles of a convex quadrilateral is 360°? _____

   _____

Geometry                                                   Student Technology Guide   **57**

NAME _____ CLASS _____ DATE _____

# Student Technology Guide
## 9.4 Angles Formed by Secants and Tangents

Star shapes appear in comic books, advertisements, banners, and so on. Using your knowledge of circles and tangents, you can study them mathematically. The diagram at the right shows an irregular star constructed by lines tangent to a given circle.

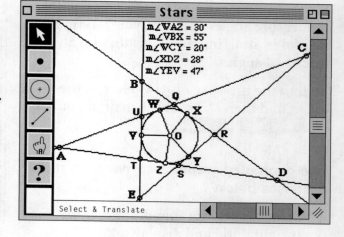

**Use geometry graphics software in the exercises below.**

1. a. Sketch a circle and place five points on it in such a way that no two of them lie along a diameter. At these points, sketch radii.
   b. At each of the five points, sketch lines perpendicular to the radii.
   c. Mark the points of intersection of the tangents sketched in part **b**. Label points *A*, *B*, *C*, *D*, and *E* as shown.
   d. Find and display m∠WAZ, m∠VBX, m∠WCY, m∠XDZ, and m∠YEV. Then calculate the sum of the measures.

2. a. Drag the points around the circle so that all the points where the tangents intersect are visible on your computer screen.
   b. Write a conjecture about the sum of the angle measures of the five points of a five-pointed star. Test your conjecture by experimenting with the diagram from Exercise 1.

_____

3. Use one of your theorems from Lesson 9.4 to write out the sum of the angle measures at points *A*, *B*, *C*, *D*, and *E*.

_____

4. Write and test a conjecture about m∠AUB + m∠BQC + m∠CRD + m∠DSE + m∠ETA.

_____

5. Repeat Exercises 1 and 2 with six points on a circle. Write conjectures similar to those you wrote in Exercises 2 and 4. Describe any patterns.

_____

_____

**58**   Student Technology Guide                                    Geometry

NAME _____ CLASS _____ DATE _____

# Student Technology Guide
## 9.5 Segments of Tangents, Secants, and Chords

In Lesson 9.5, you learned the following fact about intersecting chords:

$$AE \cdot EB = CE \cdot ED$$

This fact is used frequently to solve problems involving chords.

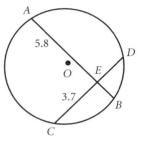

**Example:** In $\odot O$ at right, $AB = x$, $CD = 0.8x$, $AE = 5.8$, and $EC = 3.7$. Find $ED$ and $EB$. Round your answers to the nearest tenth of a unit.

- Write an equation: $5.8(x - 5.8) = 3.7(0.8x - 3.7)$.

$$x = \frac{(5.8)^2 - (3.7)^2}{5.8 - 3.7(0.8)}$$

- Press ( 5.8 $x^2$ − 3.7 $x^2$ ) ÷
  ( 5.8 − 3.7 × .8 ) ENTER

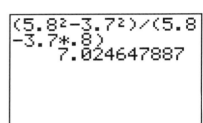

From the display, $x \approx 7.0$. Thus, $CE \approx 7.0 - 3.7 \approx 3.1$ and $EB \approx 7.0 - 5.8 \approx 1.2$.

In $\odot M$, chords $\overline{PQ}$ and $\overline{RS}$ intersect at point $Z$. Use a calculator to find the specified length. Round your answers to the nearest tenth.

1. $PZ = 12.5$, $ZQ = 4.2$, $RZ = 9.8$, $ZS =$ _____
2. $PQ = 15.5$, $PZ = 10.4$, $ZS = 8.6$, $RZ =$ _____
3. $PQ = RS$, $PZ = 3.7$, $RZ = 2.9$, $ZS =$ _____
4. $RS = 20$, $ZS = ZQ = 6.3$, $PZ =$ _____

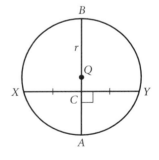

In the diagram at the right, $\overline{AB}$ is a diameter of $\odot Q$ with radius $r$ and $\overline{XY}$ is a chord perpendicular to $\overline{AB}$ at point $C$. From the diagram:

$$(XC)^2 = (YC)^2 = AC \cdot BC = AC(2r - AC)$$
$$XC = YC = \sqrt{AC(2r - AC)}$$

If $\odot Q$ has a radius of 10 inches and point $C$ is inside the circle 3 inches from the circle, then you can find $XC$ and $CY$ by using the key sequence below.

2nd $x^2$ 3 ( 2 × 10 − 3 ) ) ENTER

Refer to $\odot Q$. In Exercises 5–7, write an equation to find $XC$ and $YC$. Then use a calculator to find them. Round your answers to the nearest tenth.

5. $r = 24$; $AC = 10$     6. $r = 30$; $AC = 8.5$     7. $r = 100$; $AC = 100$

Geometry                                    Student Technology Guide

# Student Technology Guide
## 9.6 Circles in the Coordinate Plane, page 1

To graph a function on a graphics calculator, you enter the expression that defines the function. The equation $x^2 + y^2 = r^2$ defines a circle, but $y$ is not a function of $x$. To graph a circle given its equation, you must write two equations that define $y$.

$$x^2 + y^2 = 25$$
$$y^2 = 25 - x^2$$
$$y = \pm\sqrt{25 - x^2}$$

**Example:** To graph $x^2 + y^2 = 25$, you need to carry out three steps.
- Write and enter two expressions, $\sqrt{25 - x^2}$ and $-\sqrt{25 - x^2}$.

  [Y=] [2nd] [$x^2$] 25 [-] [X,T,$\theta$,n] [$x^2$] [ ) ] [ENTER] [(-)] [2nd] [$x^2$]
  25 [-] [X,T,$\theta$,n] [$x^2$] [ ) ] [ENTER]

- Select window settings that give a true circle. An $x$-range that is about 1.5 times as long as the $y$-range will accomplish this. On the display below, $-9 \le x \le 9$ and $-6 \le y \le 6$.
- Press [GRAPH].

The required displays are shown below.

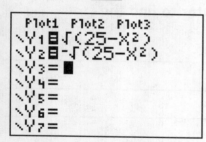
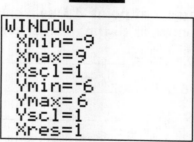
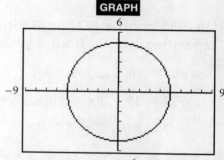

Using the example above as a guide, write each equation as a pair of equations for $y$ in terms of $x$, and then graph the equations.

1. $x^2 + y^2 = 16$ _____  2. $x^2 + y^2 = 36$ _____
3. $x^2 + y^2 = 49$ _____  4. $x^2 + y^2 = 64$ _____
5. $x^2 + y^2 = 81$ _____  6. $x^2 + y^2 = 100$ _____
7. $(x - 1)^2 + (y - 2)^2 = 3^2$ _____
8. $(x + 1)^2 + (y + 2)^2 = 3^2$ _____
9. $(x - 1)^2 + (y + 2)^2 = 3^2$ _____

10. a. Graph $(x - 2)^2 + (y - 1)^2 = 4$ and $(x - 4)^2 + (y - 1)^2 = 1$ on the same calculator display.
    b. At how many points do the graphs intersect? _____
    c. Use [Trace] to approximate the coordinates of the points of intersection. _____

NAME _____ CLASS _____ DATE _____

# Student Technology Guide
## 9.6 Circles in the Coordinate Plane, page 2

To graph an equation such as $(x - 2)^2 + (y - 1)^2 = 4$ by using geometry graphics software, you take an approach different from the approach you would take if you were using a graphics calculator. The following steps illustrate how to use the software for the graphing:

- From the main menu, select Graph Show Grid.
- From the equation, read the radius and coordinates of the center.

  radius: $\sqrt{4} = 2$; center: $C(2, 1)$
- Plot the center and a point on the circle, such as $D(2 + 2, 1) = D(4, 1)$.
- Select the center $C$ and point $D$. From the main menu, select Construct Circle By Center + Point.

The diagram at the right above shows the construction.

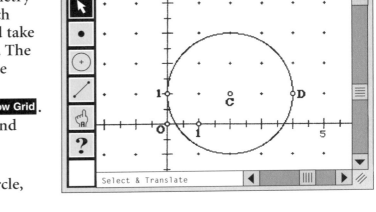

**Use geometry graphics software for the following exercises:**

11. a. Using the procedure outlined above, graph $(x - 2)^2 + (y - 1)^2 = 4$.
    b. On the same sketch, graph $(x - 4)^2 + (y - 1)^2 = 1$.
    c. At how many points do the graphs intersect? _____

To find the coordinates of any points where two circles intersect, follow these steps:
- Select the two circles.
- From the main menu, select Construct Point At Intersection.
- Select the points of intersection.
- From the main menu, select Measure Coordinates.

12. The diagram at right shows two circles that intersect. However, you cannot accurately read the coordinates of the points of intersection.
    a. Write the information needed to sketch each circle with geometry graphics software. _____
    _____
    b. Sketch the two circles and then locate and mark the points of intersection.
    c. Approximate the coordinates of the points of intersection.
    _____

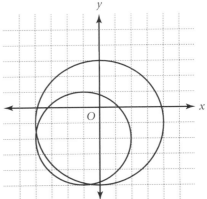

Geometry  Student Technology Guide

NAME _____ CLASS _____ DATE _____

# Student Technology Guide
## 10.1 Tangent Ratios

The graphics calculator screen at right shows the MODE display. Notice that the word *degree* is highlighted in the third line. When you begin to solve a problem involving angles, check the mode screen to be sure that *degree* is the angle measurement system you're using.

Below are simple examples to show you how to work with the tangent ratio.

**Example:** Find tan 42.5°.
  Check that the calculator is in degree mode.
  • Press TAN 42.5 ) ENTER.
  From the display, tan 42.5° ≈ 0.9163.

**Example:** Find m∠A if tan A ≈ 0.6225.
  Check that the calculator is in degree mode.
  • Press 2nd TAN .6225 ) ENTER.
  From the display, m∠A ≈ 31.9°.

**Find the tangent of each angle. Give each value to the nearest ten-thousandth.**

1. tan 45° _____    2. tan 72.6° _____    3. tan 22.4° _____    4. tan 55.8° _____

5. tan 80° _____    6. tan 10° _____    7. tan 30° _____    8. tan 60° _____

**Find the measure of each angle. Give each angle measure to the nearest tenth of a degree.**

9. tan A ≈ 0.1256    10. tan A ≈ 0.8825    11. tan A ≈ 1.3567    12. tan A ≈ 0.1111

_____    _____    _____    _____

13. tan A = 1    14. tan A ≈ 0.4141    15. tan A ≈ 1.2247    16. tan A ≈ 0.1378

_____    _____    _____    _____

17. Experiment to test the truth of the following statement:
    If 0 ≤ tan A ≤ 1, then 0° ≤ m∠A ≤ 45°. _____

_____

62    Student Technology Guide    Geometry

NAME _____ CLASS _____ DATE _____

# Student Technology Guide
## 10.2 Sines and Cosines

Trigonometry has many applications, including triangle problems. One such problem is stated below and illustrated at right.

Jasmine constructed a vertical tower 18 feet tall and attached support wires $\overline{AC}$ and $\overline{AD}$ to hold it in place. After attaching the wires, she realized that she needed those same lengths of wire for another project. She used a large protractor to find the measures of the angles that the wires make with the ground. To the nearest tenth of a foot, how long is each support wire?

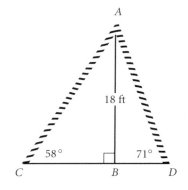

**Example:** Use a calculator to answer the question stated above. Set the calculator in degree mode.

Write two equations: $\frac{18}{\sin 58°} = AC$ and $\frac{18}{\sin 71°} = AD$.

- To find $AC$, press 18 ÷ SIN 58 ) ENTER.
- To find $AD$, press 18 ÷ SIN 71 ) ENTER.

The wires are about 21.2 and 19.0 feet long.

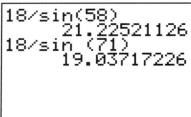

**Find each length. Round your answers to the nearest tenth of a unit.**

1. $KM$ and $LM$

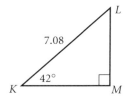

2. $YW$, $XW$, and $WZ$

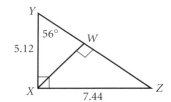

3. $PS$ and $QR$

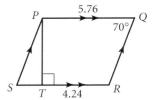

To find the measure of an angle in a right triangle given the lengths of two sides, use inverses. For example, to find m∠$BDC$ using $BC$ and $DC$, press 2nd TAN 4.16 ÷ 5.36 ) ENTER.

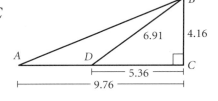

**Find the measure of each angle to the nearest tenth of a degree.**

4. ∠$BDC$ _____   5. ∠$BAC$ _____   6. ∠$DBC$ _____

7. ∠$ADB$ _____   8. ∠$ABC$ _____   9. ∠$ABD$ _____

10. Find m∠$BDC$ by using a key sequence different from the one given.

_____

Geometry            Student Technology Guide

# Student Technology Guide
## 10.4 The Law of Sines

Mindy and Jack were presented with the information about $\triangle ABC$ given at right. They want to find $m\angle C$, $BC$, and $m\angle A$. To do this, they know to use the law of sines.

$m\angle B = 35°$
$AB = 9$
$AC = 7$

**Example:** Write the Law of Sines with given information inserted:
$\frac{\sin C}{9} = \frac{\sin 35°}{7}$. Then $\sin C = \frac{9 \sin 35°}{7}$.

- Check that the calculator is in degree mode.
- Press 9 [×] [SIN] 35 [)] [÷] 7 [ENTER].

Since $\sin C > 0$, there are two possible triangles.

- Find $m\angle C$. Press [2nd] [SIN] [2nd] [(-)] [)] [ENTER].

Solution 1: $m\angle C \approx 47.5°$. Then use this to find that $BC \approx 12.1$ and $m\angle A \approx 97.5°$.

Solution 2: You know that $m\angle B = 35°$, $AB = 9$, $AC = 7$, $m\angle C \approx 132.5°$, and $m\angle A \approx 12.5°$. The display at right shows one way to find the second possibility for $BC$.

```
9*sin(35)/7
       .7374554182
sin⁻¹(Ans)
        47.51510498
7*sin(12.5)/sin(35)
        2.64145666
```

For each set of data, use a calculator to find out whether there is one possible triangle, two possible triangles, or no possible triangle. Use the figure at the right for reference.

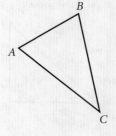

1. $m\angle A = 60°$, $m\angle B = 45°$, $BC = 6$ _____
2. $BC = 14$, $AC = 18$, $m\angle A = 38.2°$ _____
3. $m\angle A = 30°$, $BC = 25$, $AC = 20$ _____
4. $AB = BC = 10$, $m\angle A = 120°$ _____

For Exercises 5–7, consider isosceles $\triangle XYZ$ in which $\overline{XY} \cong \overline{YZ}$. Find each angle measure to the nearest tenth of a degree and the length of the base to the nearest tenth of a unit.

5. $XY = YZ = 10$, $m\angle X = 70°$ _____
6. $XY = YZ = 20$, $m\angle Z = 20°$ _____
7. $XY = YZ = 25$, $m\angle Y = 70°$ _____

8. Does the information $KL = 36$, $LM = 12$, and $m\angle K = 80°$ determine a triangle? Explain your response.

_____

NAME _____ CLASS _____ DATE _____

# Student Technology Guide
## 10.5 The Law of Cosines

In Lesson 10.5, you learned the law of cosines.
$a^2 = b^2 + c^2 - 2bc \cos A \quad b^2 = a^2 + c^2 - 2ac \cos B \quad c^2 = a^2 + b^2 - 2ab \cos C$

The following example shows how to apply the law of cosines to find the angle measures in a triangle given the lengths of the three sides.

**Example:** In $\triangle DEF$, $d = 7$, $e = 15$, and $f = 14$. Find m$\angle D$, m$\angle E$, and m$\angle F$.

To find m$\angle D$, write m$\angle D = \cos^{-1}\left(\dfrac{d^2 - (e^2 + f^2)}{-2ef}\right)$.

- Press [2nd] [COS] [ ( ] 7 [$x^2$] [ − ] [ ( ] 15 [$x^2$] [ + ] 14 [$x^2$] [ ) ] [ ) ] [ ÷ ] [ ( ] [(−)] 2 [×] 15 [×] 14 [ ) ] [ENTER].

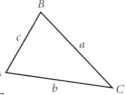

Thus, m$\angle D \approx 28°$.

- To find m$\angle E$, you edit the expression already entered. To do this, press [2nd] [ENTER]. Use [◄] and [►] to highlight the numbers you need to change. Enter the new numbers and press [ENTER].

- Follow the process just described to find m$\angle E \approx 84°$ and m$\angle F \approx 68°$.

**Find the measures of the angles in each triangle.**

1. In $\triangle ABC$, $a = 100$, $b = 100$, and $c = 150$. _____

2. In $\triangle ABC$, $a = 12$, $b = 18$, and $c = 27$. _____

3. In $\triangle ABC$, $a = 25.5$, $b = 30.6$, and $c = 40$. _____

4. In $\triangle ABC$, $a = 10$, $b = 10$, and $c = 5$. _____

You can also use the law of cosines to find the area of a triangle. The area of $\triangle DEF$ is given by $\dfrac{1}{2}ef \sin D$, as the display at right shows. After calculating m$\angle D$, press .5 [×] 15 [×] 14 [×] [SIN] [2nd] [(−)] [ ) ] [ENTER] to get 48.7 square units.

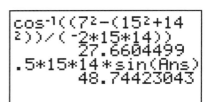

**Find the area of each triangle.**

5. In $\triangle ABC$, $a = 10$, $b = 12$, and $c = 15$. Find m$\angle A$ and use it to find the area of $\triangle ABC$.

_____

6. In $\triangle ABC$, $a = 4$, $b = 14$, and $c = 15$. Find m$\angle A$ and use it to find the area of $\triangle ABC$.

_____

Geometry            Student Technology Guide    65

NAME _____ CLASS _____ DATE _____

# Student Technology Guide
## 10.6 Vectors in Geometry, Page 1

A vector is a quantity determined by magnitude and direction. In the sketch at right, you can see two vectors. $\overrightarrow{OP}$ has a magnitude of 1.70 inches and makes an 11° angle with $\overrightarrow{OZ}$. The second vector, $\overrightarrow{OQ}$, has a magnitude of 1.20 inches and makes a 56° angle with $\overrightarrow{OZ}$. The sketch also shows a parallelogram, OPRQ, whose diagonal $\overline{OR}$ represents the sum of vectors $\overrightarrow{OP}$ and $\overrightarrow{OQ}$. (Note: When points O and P designate a vector, the common notation is $\overrightarrow{OP}$. However, geometry graphics software may not allow for a pictorial representation of this.)

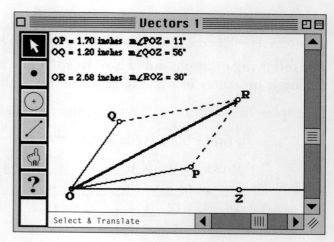

**Use geometry graphics software in the following exercises:**

1. Follow the steps below to find the magnitude and direction of the vector sum $\overline{OR}$ formed by $\overline{OP}$ and $\overline{OQ}$.
   a. Open a new sketch. Sketch $\overrightarrow{OZ}$.
   b. Sketch $\overline{OP}$, which is 1.70 inches long and makes an 11° angle with $\overrightarrow{OZ}$.
   c. Sketch $\overline{OQ}$, which is 1.20 inches long and makes a 56° angle with $\overrightarrow{OZ}$.
   d. Sketch $\overline{QR}$ the same length as $\overline{OP}$ and parallel to $\overline{OP}$. Then sketch $\overline{PR}$ the same length as $\overline{OQ}$ and parallel to $\overline{OQ}$.
   e. Sketch $\overline{QR}$. Select O and R. From the main menu, select **Measure Distance**. Then select R, O, and Z. From the main menu, select **Measure Angle**.

Follow the method from Exercise 1 to find the magnitude and direction of each vector sum. Give lengths to the nearest hundredth of a unit and angle measures to the nearest tenth of a degree.

2. $\overline{OP}$ is 2 inches long and m∠POZ = 25°.
   $\overline{OQ}$ is 2 inches long and m∠QOZ = 65°.

   magnitude: _____

   direction: _____

3. $\overline{OP}$ is 2.4 inches long and m∠POZ = 15°.
   $\overline{OQ}$ is 1.5 inches long and m∠QOZ = 50°.

   magnitude: _____

   direction: _____

NAME _____ CLASS _____ DATE _____

 **Student Technology Guide**
*10.6* Vectors in Geometry, Page 2

When you place an arrow on the coordinate plane with its tail at the origin, you can represent the arrow with an ordered pair. In the sketch at right, the arrow with its tail at the origin and tip at $P(10, 2)$ is represented by $(10, 2)$, and the arrow with its tail at the origin and its tip at $Q(4, 6)$ is represented by $(4, 6)$.

You can use the grid feature to investigate vector addition with coordinates.

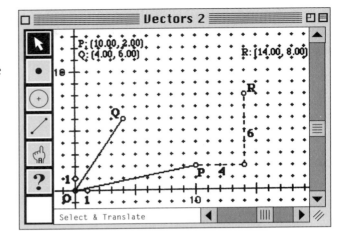

**Use geometry graphics software in the following exercises:**

4. Follow the steps below to find the coordinates of the vector sum $\overline{OR}$ determined by $\overline{OP}$ and $\overline{OQ}$.
   a. Open a new sketch. From the main menu, select **Graph** **Show Grid**.
   b. Place points at $P(10, 2)$ and $Q(4, 6)$. Sketch $\overline{OP}$ and $\overline{OQ}$.
   c. Translate point $P$ to the right 4 units and up 6 units. Label the image of the translation as point $R$. Its coordinates are the coordinates of the vector sum. Display them by selecting $R$ and **Measure** **Coordinates**.

**Use the software to find the coordinates of each vector sum described below.**

5. $P(3, 1)$ and $Q(6, 3)$ _____

6. $P(5, 0)$ and $Q(0, 5)$ _____

7. $P(-3, 2)$ and $Q(6, 1)$ _____

8. $P(-4, -3)$ and $Q(6, 1)$ _____

9. $P(-3, 0)$ and $Q(3, 0)$ _____

10. $P(-4, -3)$ and $Q(4, 3)$ _____

11. a. Based on your work in Exercises 5–10, write a rule that will give the coordinates of the sum of two vectors whose tips are represented by $(a, b)$ and $(c, d)$.

    _____

    b. Apply your rule from part a to find the sum of the vectors whose tips are represented by $P(4.6, -2.7)$ and $Q(10.3, 9.5)$. _____

    c. What can you say about two vectors whose sum has the coordinates $(0, 0)$? _____

Geometry                                      Student Technology Guide    **67**

NAME _____ CLASS _____ DATE _____

# Student Technology Guide
## 10.7 Rotations in the Coordinate Plane

Many graphics calculators allow you to enter matrices and perform operations on them. To enter a matrix into a graphics calculator, you need to name the matrix, state its dimensions, and then enter its entries in order row by row. To enter $\begin{bmatrix} -2 & 0 \\ 3 & 5 \end{bmatrix}$, press

[MATRX] [EDIT] [1:[A]] [ENTER] 2 [ENTER] 2 [ENTER] [(-)] 2 [ENTER] 0 [ENTER] 3 [ENTER] 5 [ENTER].

**Example:** Let $A = R_{45} = \begin{bmatrix} \cos 45° & -\sin 45° \\ \sin 45° & \cos 45° \end{bmatrix}$ and $B = \begin{bmatrix} 0 & 2 & 5 \\ 0 & 4 & 3 \end{bmatrix}$.

Find the coordinates of the image of the triangle represented by matrix $B$ under the rotation represented by matrix $A$.

- Enter matrices $A$ and $B$ as explained above.
- Press [MATRX] [NAMES] [1:[A]] [×] [MATRX] [NAMES] [2:[B]] [ENTER] [ENTER].
- To see the rest of the matrix press [▶].

```
[[A]*[B]
[[0  -1.41421356...
 [0  4.242640687...
```

**Refer to the rotation matrices at right. Find each product listed below. On the coordinate grid provided, graph the image.**

$R_{90} = \begin{bmatrix} \cos 90° & -\sin 90° \\ \sin 90° & \cos 90° \end{bmatrix}$  $R_{180} = \begin{bmatrix} \cos 180° & -\sin 180° \\ \sin 180° & \cos 180° \end{bmatrix}$

$R_{-90} = \begin{bmatrix} \cos (-90°) & -\sin (-90°) \\ \sin (-90°) & \cos (-90°) \end{bmatrix}$  $R_{270} = \begin{bmatrix} \cos 270° & -\sin 270° \\ \sin 270° & \cos 270° \end{bmatrix}$

1. $R_{-90}B$ _____

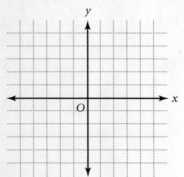

2. $R_{270}B$ _____

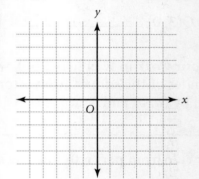

3. The expression $(R_\theta)^n B$, where $n$ is a positive integer, represents the image after $n$ successive rotations. Find $(R_{45})^8 B$. Without actually calculating the product, what can you say about $(R_{45})^{(8n)}B$? _____

NAME _____ CLASS _____ DATE _____

# Student Technology Guide
## 11.3 Graph Theory

The diagram at the left below represents a group of five people in a network who communicate with one another in some way. For example, $P$ and $Q$ can communicate directly with each other. Person $S$ can communicate directly with $R$, but $R$ cannot communicate directly with $S$.

Matrix $A$ represents communication lines among the members. An entry of 1 indicates direct communication from one member to another. An 0 entry indicates no direct line of communication between them.

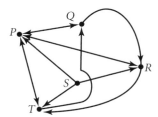

$$A = \begin{array}{c} \\ P \\ Q \\ R \\ S \\ T \end{array} \begin{array}{c} \begin{array}{ccccc} P & Q & R & S & T \end{array} \\ \left[\begin{array}{ccccc} 0 & 1 & 1 & 0 & 1 \\ 1 & 0 & 1 & 0 & 0 \\ 1 & 0 & 0 & 0 & 1 \\ 1 & 0 & 1 & 0 & 1 \\ 1 & 1 & 0 & 0 & 0 \end{array}\right] \end{array}$$

A power of matrix $A$ reveals which members can communicate with one another via intermediary members.

To find $A^2$, follow the example below.

**Example:** • Enter the entries in matrix $A$. Press MATRX EDIT
1:[A] ENTER 5 ENTER 5 ENTER 0 ENTER 1 ENTER 1 ENTER ...
0 ENTER 0 ENTER
• Then press MATRX NAMES 1:[A] $x^2$ ENTER.
The display at the right shows the result.

```
[A]²
    [[3 1 1 0 1]
     [1 1 1 0 2]
     [1 2 1 0 1]
     [2 2 1 0 2]
     [1 1 2 0 1]]
```

**Use a graphics calculator as needed.**

1. **a.** How many communication paths are there from member $Q$ to member $T$ via one group member? Compare this number with the entry in row 2 and column 5 of $A^2$. _____
   **b.** What do you think $A^2$ tells you about the network?
   _____

2. **a.** Find $A^3$. (Use ^ 3 rather than $x^2$ in the key sequence above.) _____
   **b.** How many communication paths are there from member $R$ to member $P$ via two group members? (You may enter and exit $P$ along the way if you wish.) Compare this number with the entry in row 3 and column 1 of $A^3$. _____
   **c.** What do you think $A^3$ tells you about the network?
   _____

Geometry                                    Student Technology Guide

NAME _____ CLASS _____ DATE _____

# Student Technology Guide
## 11.4 Topology: Twisted Geometry, page 1

The diagram at right shows a horizontal line segment. Vertical arrows of differing lengths show points that determine a curved, or bent, line segment. Using a function, you can deform a line segment to get a topologically equivalent bent line segment.

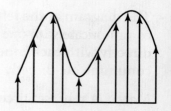

The graphics calculator displays below show what happens when some simple functions of $x$ are applied to the interval $-4 \leq x \leq 4$. Each curve is topologically equivalent to a line segment.

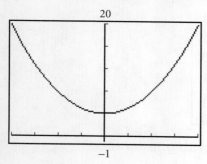
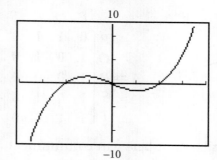
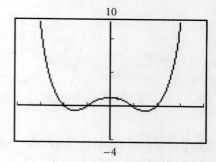

In Exercises 1–6, enter each function into the function list, enter $-4 \leq x \leq 4$ into the window settings, and then graph the function. Describe how the function bends the line segment defined by $-4 \leq x \leq 4$.

1. $y = 2x + 1$

2. $y = |x| + 1$ (Press MATH, select NUM 1:abs( )

3. $y = 2x^2 - 1$

4. $y = 0.25(x + 2)(x)(x + 2)$

5. $y = \sqrt{16 - x^2}$

6. $y = 0.01(x + 4)(x + 2)(x - 2)(x - 4)$

When you choose a function whose domain is an interval and whose graph consists of more than one piece, the new figure is not topologically equivalent to the original.

7. Let $-4 \leq x \leq 4$ and $y = \dfrac{1}{(x + 2)(x - 2)}$. Graph this function. What effect does the function have on the interval? _____

70  Student Technology Guide  Geometry

NAME _____ CLASS _____ DATE _____

# Student Technology Guide
## 11.4 Topology: Twisted Geometry, page 2

You can use geometry graphics software to perform many actions that deform one figure into another that is not topologically equivalent to the first.

Notice that in the sketch at right, a simple closed figure with vertices $A, B, C, \ldots, J$, and $K$ has been copied and pasted to a new location. Then various vertices were dragged to turn the original figure into a figure eight. This figure is not topologically equivalent to the original because the figure eight has two interior parts.

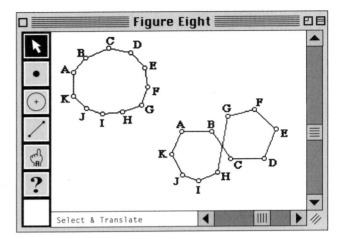

**Use geometry graphics software.**

8. a. Open a new sketch. Draw a convex polygon that has 20 vertices.
   b. Drag vertices until you get a figure eight with an extra loop, a figure with three interior parts.
   c. Why are the two figures not topologically equivalent?

   _____

   _____

All convex polygons are topologically equivalent to each other and to circles.

9. a. Open a new sketch. Draw a convex polygon that has 16 vertices.
   b. Transform the original polygon into a convex octagon. Explain how this action maintains the essential topological nature of the original polygon.

   _____

   _____

   c. Using your octagon from part b, drag vertices to form a quadrilateral. Explain how that action maintains the essential topological nature of the original polygon.

   _____

10. Devise a way to turn a quadrilateral into an octagon and then into a dodecagon. _____

   _____

   _____

**Geometry**  **Student Technology Guide**

NAME _____ CLASS _____ DATE _____

# Student Technology Guide
## 11.6 Fractal Geometry, page 1

The four diagrams below illustrate the first four stages in the construction of the Sierpinski triangle, also called the Sierpinski gasket, based on an initial equilateral triangle. As the construction process continues, the higher the stage number, the more "holes" in the triangle and the more boundary lines around the holes.

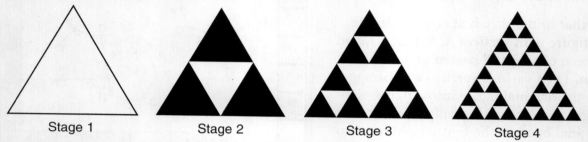

Stage 1   Stage 2   Stage 3   Stage 4

Suppose that at stage 1 you have an equilateral triangle with sides of 1 unit. Let $n$ represent the stage number, $s$ represent the number of shaded triangles at stage $n$, $L$ represent the length of each shaded triangle at stage $n$, and $P$ represent the perimeter of all the shaded triangles at stage $n$.

**In Exercises 1–3, write a formula for the specified variable in terms of $n$.**

1. $s$

2. $L$

3. $P$

4. Using your formulas from Exercises 1–3, create a spreadsheet like the one shown at right.

5. What can you conclude about $P$ as $n$ increases?

6. a. The area $A$ of an equilateral triangle with sides of $L$ units is approximated by $A = 0.4330L^2$. Modify your spreadsheet to find the total area of all shaded triangles at stage $n$.
   b. What can you say about the total area of the shaded triangles as $n$ increases?

7. a. Modify your spreadsheet to find the total area of all unshaded triangles at stage $n$.
   b. What can you say about the total area of the unshaded triangles as $n$ increases?

# Student Technology Guide
## 11.6 Fractal Geometry, page 2

Now suppose that you begin with a square with sides of 1 unit. Below are the first four stages in the process of removing smaller squares, each of which is one-third as long on a side as in the preceding stage.

Stage 1

Stage 2

Stage 3

Stage 4

Let $S_n$ represent the area of the shaded regions at stage $n$. $S_1$, $S_2$, $S_3$, and $S_4$, shown at right, will represent the area of the shaded regions at stages 1, 2, 3, and 4, respectively.

$S_1 = 1$

$S_2 = 1 - \left(\frac{1}{3}\right)^2$

$S_3 = 1 - \left(\frac{1}{3}\right)^2 - 8\left(\left(\frac{1}{3}\right)\left(\frac{1}{3}\right)\right)^2$

$S_4 = 1 - \left(\frac{1}{3}\right)^2 - 8\left(\left(\frac{1}{3}\right)\left(\frac{1}{3}\right)\right)^2 - 64\left(\left(\frac{1}{3}\right)\left(\frac{1}{3}\right)\left(\frac{1}{3}\right)\right)^2$

**For Exercises 8–10, refer to the diagrams and sums shown here.**

8. Using $S_4$, write an expression for $S_5$.

___

9. a. Explain how to write an expression for $S_n$.

___

___

b. Write an expression for $S_n$ in terms of $S_{n-1}$.

___

Each expression for $S_n$ is called a partial sum. If $S_n$ is in a cell of a spreadsheet, then the cell below it contains $S_{n+1}$ and is the value of $S_n$ with a certain amount subtracted from it.

10. a. Set up a spreadsheet like the one at right.
    b. Use spreadsheet formulas and your algebraic formula from part b of Exercise 9 to complete column B.
    c. How does $S_n$ change as $n$ increases?

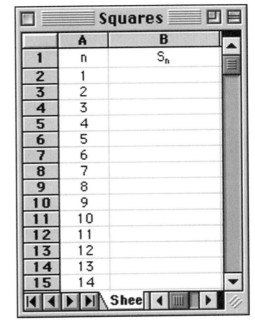

___

___

Geometry

# Student Technology Guide
## 11.7 Other Transformations: Projective Geometry

The dilation $T(x) = kx$ and $T(y) = ky$, where $k > 0$, can be represented by the matrix $\begin{bmatrix} k & 0 \\ 0 & k \end{bmatrix}$. If you represent square $OABC$ as the matrix $\begin{bmatrix} 0 & 0 & 4 & 4 \\ 0 & 4 & 4 & 0 \end{bmatrix}$, then the product $\begin{bmatrix} k & 0 \\ 0 & k \end{bmatrix} \begin{bmatrix} 0 & 0 & 4 & 4 \\ 0 & 4 & 4 & 0 \end{bmatrix}$ gives the coordinates of the enlarged or reduced square.

The transformation $T(x) = mx$ and $T(y) = ny$, where $m > 0$, $n > 0$ and $m \neq n$, has an unusual effect on square $OABC$.

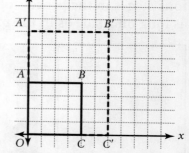

**Example:** Let $A = \begin{bmatrix} 1.5 & 0 \\ 0 & 2 \end{bmatrix}$ and $B = \begin{bmatrix} 0 & 0 & 4 & 4 \\ 0 & 4 & 4 & 0 \end{bmatrix}$.

Find $AB$. Plot the vertices represented by $AB$.
- Enter matrix $A$. Press MATRX EDIT 1:[A] ENTER 2 ENTER 2 ENTER 1.5 ENTER 0 ENTER 0 ENTER 2 ENTER.
- In similar fashion, enter matrix $B$.
- Press MATRX NAMES 1:[A] × MATRX NAMES 2:[B] ENTER.

The image $OA'B'C'$ of square $OABC$ is shown on the grid above, along with square $OABC$. Figure $OA'B'C'$ is a rectangle.

**Use a calculator in the exercises below.**

1. Suppose that you want the image of square $OABC$ to be a rectangle whose horizontal side is greater than its vertical side.
   a. How would you modify matrix $A$ to accomplish this? _____
   b. Write the new matrix. _____
   c. Find $AB$ to verify your responses in parts a and b. _____

Let $C = \begin{bmatrix} 1 & 2 \\ 0 & 1 \end{bmatrix}$ and $D = \begin{bmatrix} 1 & 0 \\ 2 & 1 \end{bmatrix}$. Find each product.

**Graph each image at the right. Describe the effect of $C$ and $D$.**

2. $CB$ _____   3. $DB$ _____

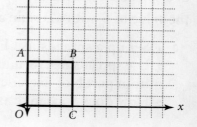

4. Let $E = \begin{bmatrix} 2 & -1.5 \\ 0 & 1 \end{bmatrix}$. Describe the polygon represented by $EB$.
_____

74  Student Technology Guide  Geometry

NAME _____ CLASS _____ DATE _____

# Student Technology Guide
## 12.2 And, Or, and Not in Logical Arguments, page 1

At right are spreadsheet truth tables for $p$ AND $q$ and $p$ OR $q$. To create the spreadsheet, enter the T/F combinations for $p$ and $q$ in columns A and B. Represent T by 1 and F by 0. To complete the truth table for $p$ AND $q$ and $p$ OR $q$, enter the formulas below into cells C2 and D2, respectively. Then FILL DOWN.
=IF(AND(A2=1,B2=1),1,0)
=IF(OR(A2=1,B2=1),1,0)

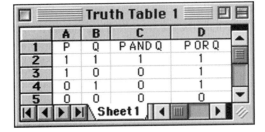

To show negation of a statement, use NOT( ). For example, if A1 contains 1 and B1 contains NOT(A1), then B1 will show 0.

**In Exercises 1–4, use a spreadsheet to show that each pair of expressions, abbreviated EXP 1 and EXP 2, is logically equivalent. Fill in each spreadsheet with what is on your computer display.**

1. EXP 1: $\sim(p$ OR $q)$; EXP 2: $\sim p$ AND $\sim q$

2. EXP 1: $\sim(\sim p$ OR $\sim q)$; EXP 2: $p$ AND $q$

3. EXP 1: $\sim(\sim p$ AND $\sim q)$; EXP 2: $p$ OR $q$

4. EXP 1: $p$ OR $(p$ AND $q)$; EXP 2: $p$ OR $q$

5. Suppose that you are given $\sim(\sim\sim p$ or $\sim\sim q)$. Write a simpler expression that you think is logically equivalent to the given one. Then use a spreadsheet to verify your simplification. Enter your results into the spreadsheet display at right.

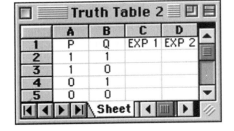

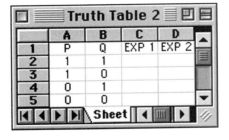

Geometry                            Student Technology Guide

# Student Technology Guide

## 12.2 And, Or, and Not in Logical Arguments, page 2

To explore the logical equivalence of more complicated expressions such as $p$ AND ($q$ OR $r$) and ($p$ AND $q$) OR ($p$ AND $q$), you will need to make spreadsheet formulas using both the connectives AND and OR. You will also need to enter all T/F combinations for $p$, $q$, and $r$ into columns A, B, and C.

**For Exercises 6–10, use a spreadsheet like the one at right.**

6. Create a spreadsheet like the one shown at right.

**Test for logical equivalence of the two given compound statements.**

7. $p$ AND ($q$ OR $r$); ($p$ AND $q$) OR ($p$ AND $r$)

8. $p$ OR ($q$ AND $r$); ($p$ OR $q$) AND ($p$ OR $r$)

9. $p$ AND ($q$ AND $r$); ($p$ AND $q$) AND $r$

10. $p$ OR ($q$ OR $r$); ($p$ OR $q$) OR $r$

# Student Technology Guide
## 12.4 Indirect Proof

A brief description of an indirect argument might go as follows:

> If you claim that a statement is true and, as a consequence, you get a statement that I can show is false, then I can say that your claim must be false.

A simple algebraic application of this reasoning is as follows:

> If I simplify an algebraic expression in $x$, expression 1, and get another algebraic expression in $x$, expression 2, then the graphs of $y =$ expression 1 and $y =$ expression 2 should coincide in the $xy$-plane. If I get two distinct graphs, then expression 1 and expression 2 are not logically equivalent.

You can use a graphics calculator to help find out if two algebraic expressions are equivalent. For example, suppose that Jenna has the expression $(x + 2)(x + 2)$ and she simplifies it to $4x^2$. You can graph $y = (x + 2)(x + 2)$ and $y = 4x^2$ and examine the graphs to see if they coincide. Since two very different graphs are visible on the graphics calculator display at right, Jenna is incorrect.

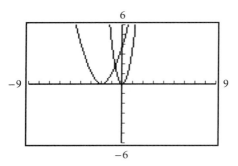

**In the following exercises:**
a. Use a graphics calculation to find out whether the two given algebraic expressions are equivalent or not equivalent.
b. If they are not equivalent, identify the error in the algebraic work that may have led to the incorrect simplification.

1. $2(x + 3)$ and $2x + 3$ _____
2. $(x + 2)(x - 2)$ and $-4x^2$ _____
3. $(x + 2)(x - 3)$ and $x^2 + 5x + 6$ _____
4. $(2x)^2(3x)$ and $6x^2$ _____

In algebra, you learned the *quadratic formula*: if $ax^2 + bx + c = 0$ and $a \neq 0$, then $x = \frac{-b \pm \sqrt{b^2 - 4ac}}{2a}$. You also learned the following fact about the graph of a *quadratic function*: the $x$-intercepts of the graph of $y = ax^2 + bx + c = 0$ are the roots of $ax^2 + bx + c = 0$.

5. When Jenna used the quadratic formula to solve $x^2 - x - 6 = 0$, she got $x = \frac{1 \pm \sqrt{-23}}{2}$ (i.e., no real roots). Use a graphics calculator to check her claim. If she made an error, what was it? _____

Geometry                                     Student Technology Guide

# Answers

## Student Technology Guide—Chapter 1

### Lesson 1.1

1–11. Check students' sketches.

### Lesson 1.2

1–2. Check students' sketches.

3. *PR* and *RQ* change as point *R* is dragged, but *PQ* and the sum *PR* + *RQ* remain constant and equal to each other; Segment Addition Postulate

4. All distances except *PQ* change as point *R* is moved. The sum *PR* + *RQ* is always greater than *PQ*. This illustrates the fact that the segment connecting two points is shorter than any other path between them.

5. infinitely many points; an ellipse

### Lesson 1.3

1–2. Check students' sketches.

3. As point *S* is dragged, the measures of angles *PQS* and *SQR* are constantly changing. The measure of ∠*PQR* and the sum "angle *PQS* + angle *SQR*" remain constant and equal to each other. This illustrates the Angle Addition Postulate.

4. Check students' sketches; the point is positioned correctly when m∠*PQS* = m∠*SQR*.

5. When point *S* is moved to the exterior of ∠*PQR*, the sum "angle *PQS* + angle *SQR*" is no longer constant and is no longer equal to the measure of ∠*PQR*. The sum is constantly changing, and it is always greater than the measure of ∠*PQR*.

6. The statement is false, but students may identify it as "sometimes true and sometimes false." See below for sample sketches.

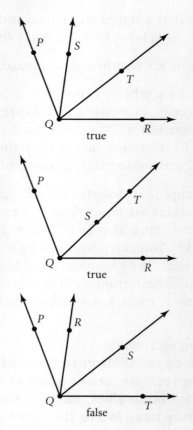

### Lesson 1.4

1. Check students' sketches.

2. Methods may vary. Sample method: Draw one segment with *A* and *C* as its endpoints and another with *C* and *B* as its endpoints. Locate the midpoint of each of these segments and label them *D* and *E*, respectively. Through *D* and *E* draw two lines perpendicular to $\overline{AB}$.

3. Check students' sketches.

4. Answers may vary. The bisectors of the four angles are always perpendicular to each other.

# Answers

5. Check students' sketches.

6. Answers may vary. In each sketch drawn with the line tool, when points $M$ and $O$ are moved, $\overline{LM}$ and $\overline{NO}$ move freely. In the sketch drawn with the construction features, it may or may not be possible to move point $M$ or point $O$; if it is, then the entire set of three lines moves with the point.

## Lesson 1.5

1. Check students' sketches.

2–6. Answers may vary. Sample responses are given.

2. The four angle bisectors form a square; bisectors of consecutive angles are perpendicular; bisectors of opposite angles are parallel.

3. They intersect at one point when $ABCD$ is a square. For all other rectangles, they behave as described in Exercise 3. When the length of $ABCD$ is twice its width, two vertices of the square formed by the angle bisectors are at midpoints of the longer sides of $ABCD$.

4. No; they form a rectangle that is not a square.

5. The angle bisectors form a non-rectangular quadrilateral. They intersect at one point for several positions of point $B$ on the same line.

6. When $ABCD$ is a rectangle, the perpendicular bisectors intersect at one point. If you drag $\overline{CD}$, the perpendicular bisectors always form a four-sided figure in which opposite sides are parallel (that is, a parallelogram). If you begin again with $ABCD$ as a rectangle and drag point $B$, the perpendicular bisectors intersect at one point whenever $\overline{AB}$ and $\overline{BC}$ are perpendicular.

## Lesson 1.6

1–2. Check students' sketches.

3. Reflections preserve size and shape.

4. Yes; regardless of a line's position, it is still a line of reflection, so size and shape are preserved.

5. The triangles continue to have the same size and shape, but the complete diagram is no longer symmetric.

6. Check students' sketches.

7. The size and shape of the triangles, the center of rotation, and the angle of rotation remain the same.

8. Check students' sketches. Answers may vary. The completed diagram will contain the original rectangle and seven copies of it, all containing one point in common. A sample result is shown below.

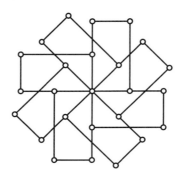

# Answers

## Lesson 1.7

1. $y = 2.5x$
   $y = 2.5(x + 3) - 2$, or $y = 2.5x + 5.5$

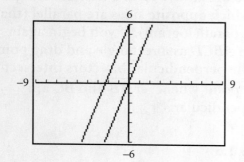

2. $y = -2x$
   $y = -2(x + 3) - 2$, or $y = -2x - 8$

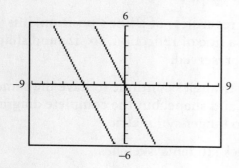

3. $y = 4x$
   $y = 4(x + 3) - 2$, or $y = 4x + 10$

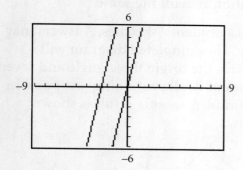

4. The new line and the original line are parallel. If both $h$ and $k$ are nonzero, then the lines will never meet.

5. $y = 2.5x$

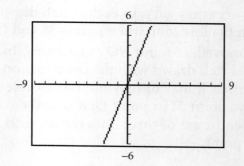

6. $y = -2x$

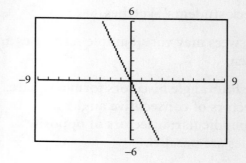

7. $y = -x$

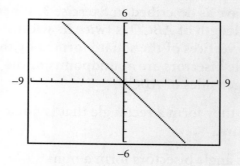

8. $y = 4x$

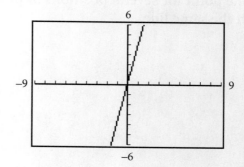

# Answers

9. The resulting line coincides with the original line because $mx$ translated to $m(-x)$ gives $-[m(-x)]$, or $mx$.

## Student Technology Guide — Chapter 2

### Lesson 2.1

1. Check students' sketches.

2. No. The perpendicular bisectors can meet in the interior or the exterior. They can even meet at a point on the triangle itself.

3. The statement is always true by the Segment Addition Postulate.

4. The statement is not always true. The sides can have the same length, but the angles are not necessarily right angles.

5. The statement is false. If the four points lie along the same line, for example, they do not determine a polygon at all.

6. $AB + BD = AD$, but $BD = BC + CD$. Therefore, by substitution, $AB + (BC + CD) = AB + BC + CD = AD$.

### Lesson 2.3

1. Check students' sketches.

2. a. Answers may vary. No; adjacent angles must have the same vertex.
   b. Answers may vary. Move point $A$ so that it coincides with point $E$. Then move point $C$ so that it lies on $\overrightarrow{ED}$. Leave point $B$ on $\overrightarrow{AB}$.

3. Answers may vary. Move point $E$ so that it coincides with point $A$. Then move points $D$ and $F$ to the interior of $\angle ABC$. Make sure that $D$ and $F$ remain distinct.

4. a. Illustrations may vary.

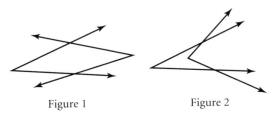

Figure 1    Figure 2

In Figure 1, the angles face, or oppose, one another. In Figure 2, the two angles face in the same direction.

b. Answers may vary. Michael's definition is not a good one for many reasons. For one thing, there could be a situation that satisfies his definition but whose meaning is opposite that of his intent, as in Figure 2. It can also happen that one angle faces to the right and the other angle faces up.

### Lesson 2.5

In Exercises 1–4, an even number is represented by $2n$ and an odd number is represented by $2n + 1$, where $n$ is an integer.

1. The product of two even numbers is always even.
   $(2k)(2n) = 2(2kn)$, which is an even number.

2. The sum of two odd numbers is always even.
   $(2k + 1) + (2n + 1) = 2(k + n + 1)$, which is an even number.

3. The square of an even number is always even because it is the product of two even numbers.
   $(2n)^2 = (2n)(2n) = 2(2n^2)$, which is an even number.

4. The square of an odd number is always odd.
   $(2n + 1)^2 = 4n^2 + 4n + 1 = 2(2n^2 + 2n) + 1$, which is an odd number.

# Answers

5. a. Check students' sketches.
   b. Check students' sketches.

6. Check students' sketches.

7. The triangles have the same size and shape.

8. The observations from Exercise 7 remain the same.

9. Choosing O as the center of rotation and 2m∠DOF as the angle of rotation, rotate △ABC counterclockwise. The image will be the same as that given by the double reflection.

10. Check students' sketches.

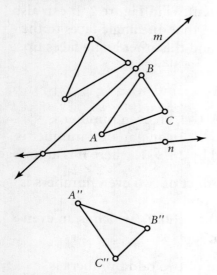

The conclusion is still true. In the diagram above, △A″B″C″ is the image of the double reflection and the image under the rotation.

There is a warning. In the conclusion from Exercise 9, the angle measure is always positive. But angles of rotation can be positive or negative depending on whether the rotation is clockwise or counterclockwise. A more accurate statement is that the angle of rotation is the measure of the angle between the lines and that the rotation angle is this measure with clockwise or counterclockwise direction.

## Student Technology Guide — Chapter 3

### Lesson 3.2

1. Check students' sketches.

2. Select $D$ and $\overline{AB}$. Construct a parallel line through $D$. Then drag $C$ so that it is also on this line. Now opposite sides are parallel, so, $ABCD$ is a parallelogram by definition.

3. To change the parallelogram into a rectangle, drag point $B$ so that m∠$ABC$ = 90°. Maintain all parallels drawn in Exercise 2.

4. Maintaining all parallels, drag vertices so that $AB = BC = CD = DA$.

5. Maintaining all parallels, drag vertices so that $AB = BC = CD = DA$ and m∠$ABC$ = 90°.

# Answers

6. The two figures below. The quadrilateral is not unique.

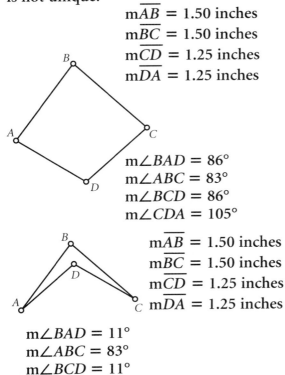

$m\overline{AB}$ = 1.50 inches
$m\overline{BC}$ = 1.50 inches
$m\overline{CD}$ = 1.25 inches
$m\overline{DA}$ = 1.25 inches

$m\angle BAD = 86°$
$m\angle ABC = 83°$
$m\angle BCD = 86°$
$m\angle CDA = 105°$

$m\overline{AB}$ = 1.50 inches
$m\overline{BC}$ = 1.50 inches
$m\overline{CD}$ = 1.25 inches
$m\overline{DA}$ = 1.25 inches

$m\angle BAD = 11°$
$m\angle ABC = 83°$
$m\angle BCD = 11°$
$m\angle CDA = 105°$

7. Check students' sketches.

8. a. Reflect $\overline{BC}$ across line *n*. Drag point C to where the image of B is.
   b. Reflect $\overline{AD}$ across line *n*. Drag point D to where the image of A is.
   c. The deformed figure has line symmetry because line *n* is the perpendicular bisector of $\overline{BC}$ and of $\overline{AD}$.

9. To achieve rotational symmetry, there must be point about which ABCD is rotated so that it coincides with itself after the rotation.

   Deform ABCD so that it is a square. Find and mark the intersection point of the diagonals, and use it as the center of rotation. The image must coincide with the original square, ABCD, and it does after rotation through a multiple of 90°.

## Lesson 3.4

1. Check students' sketches.

2. a. Check students' sketches.
   b. Check students' sketches.
   c. Answers may vary. Alternate interior angles are $\angle BEF$ and $\angle XFC$. If they have the same measure, then the lines are parallel. But $\angle BEF$ is a supplement of $\angle XEB$. Calculate and display $180° - m\angle XEB$. Drag C until $m\angle XFC$ equals the measure just calculated.

3. a. Check students' sketches.
   b. Answers may vary. Suppose that $\angle XFD$ is chosen. Then $\angle XEB$ and $\angle XFD$ are already corresponding angles. Drag C or D until the measures become equal.

4. If the two points where the transversal intersects the two lines are not used, you will not be able to apply the converse of any of the parallelism theorems.

## Lesson 3.5

1. a. $180 - (105.2 + 42.7)$
   b. $32.1°$

2. a. $180 - (60.2 + 59.4)$
   b. $60.4°$

3. $180 - (24.7 + 56.4)$; $98.9°$

4. $180 - (24.7 + 110.3)$; $45°$

5. $180 - (24.7 + 56.4) - (180 - (24.7 + 110.3))$; $53.9°$

6. $180 - (24.7 + 71.1) - 56.4$; $27.8°$

7. $180 - (180 - (24.7 + 71.1) - 56.4) - 71.1$; $81.1°$

# Answers

## Lesson 3.6

1. a. $360 = (91.5 + x + (x - 5.5) + 120.4)$
   b. $x = \frac{360 - 91.5 + 5.5 - 120.4}{2}$; 76.8

2. a. $x + (3 \times 180 - (2 \times 90 + 123 + 89.5)) = 180$
   b. $x = 180 - (3 \times 180 - (2 \times 90 + 123 + 89.5))$; 32.5

3. If $2400 = (n - 2)180$, then $n = \frac{2400}{180} + 2$; Since the answer is not a whole number, the polygon described does not exist.

## Lesson 3.7

1. Check students' sketches.

2. Answers may vary. Each successive length should be one half the length calculated in the previous step of the iteration.

3. $M_k N_k = \frac{1}{2^k} AC$

4. a. $\triangle M_k N_k B$ is also equilateral.
   b. The lengths of the sides of $\triangle M_k N_k B$ are twice as long as the sides of $\triangle M_{k+1} N_{k+1} B$.

5. Check students' sketches.

6. $R_k S_k = \frac{1}{2}(AC + M_k N_k) = \frac{1}{2}\left(AC + \frac{1}{2^n}AC\right) = \left(\frac{2^n + 1}{2^{(n+1)}}\right)AC$

## Student Technology Guide — Chapter 4

## Lesson 4.3

1. Check students' sketches.

2. $HF = IG = 1.09$ inches; $\overline{FE}$ and $\overline{DG}$ are both vertical, so they lie along parallel lines. So, $\angle FEH \cong \angle GDI$ because they are corresponding angles. $\triangle FEH \cong \triangle GDI$ by the AAS Triangle Congruence Theorem. By CPCTC $\overline{FH} \cong \overline{GI}$. So, $FH = GI$.

3. In a computer-aided sketch, congruence is much more evident than in a hand-drawn sketch which may or may not show relative sizes of things.

4. The adjustment has no effect on the problem. The information about the angle of the embankment can be considered extraneous.

5. Check students' sketches.

6. Measure $\angle ABE$ and $\angle YMH$. Use information about $AB$, $BE$, and their corresponding sides in $\triangle YMH$. That will be enough to prove that $\triangle ABE \cong \triangle YMH$ by the SAS Triangle Congruence Postulate.

7. $\overline{NA} \parallel \overline{CD}$ and $m\angle NAB = 45°$, $m\angle ABD = 45°$. $m\angle ABE = m\angle ABD + 90° = 45° + 90° = 135°$
   $\overline{N'Y} \parallel \overline{LM}$ and $m\angle ZYG = 45°$, $m\angle YGL = 45°$.
   $m\angle YMH = m\angle YGL + 90° = 45° + 90° = 135°$
   Thus, $\angle ABE \cong \angle YMH$, and $\triangle ABE \cong \triangle YMH$ by the SAS Triangle Congruence Postulate.

# Answers

8. a. m $\overline{AB}$ = 1.00 inches
m∠DBC = 45.0°
m $\overline{BC}$ = 2.00 inches

m $\overline{XY}$ = 2.00 inches
m $\overline{YZ}$ = 1.00 inches
m∠ZYI = 45.0°

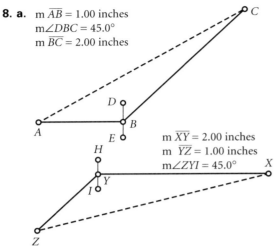

b. Using the software, measure ∠ABC and ∠ZYX. The measures are equal. The triangles are congruent by SAS, so $\overline{AC} \cong \overline{XZ}$ and AC = XZ by CPCTC. To verify that ∠ABC ≅ ∠ZYX show that both are 135° from due north.

## Lesson 4.5

1. Check students' sketches.

2. Find and display the lengths of the pairs of opposite sides. Note the equality. Alter the diagram so that ABCD is still a parallelogram, and note that the lengths remain equal.

3. Find and display the measures of the pairs of opposite sides. Note the equality. Alter the diagram so that ABCD is still a parallelogram, and note that the measures remain equal.

4. Sketch the diagonals. Locate and mark their intersection. Find and display the lengths of the four parts of the two diagonals. Point out the two equalities on the screen. Modify the diagram so that ABCD remains a parallelogram, and note that the lengths remain equal.

5. The two triangles resulting from one diagonal are congruent. To demonstrate the fact, select a congruence postulate or theorem such as SSS. Find and display the information called for in the hypothesis. Note the equalities relating corresponding parts.

6. a. Check students' sketches.
   b. rhombus
   c. The diagonals are perpendicular to one another. Find the measure of one angle resulting from the intersection of the diagonals. The measure will be 90°.

7. The claim is true. Find the measures of the other three angles. They will all be 90°. Thus, the figure has four congruent sides and four congruent angles. The figure is a square.

8. a. Check students' sketches.
   b. Sketch the diagonals. Click on one diagonal and select Construct Point at Midpoint. Do the same for the other diagonal. Label the midpoints E and F, respectively. Since the midpoints do not coincide, the diagonals cannot possibly bisect one another.

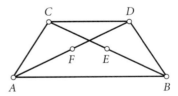

Geometry                                                                 Answers    85

# Answers

## Lesson 4.6

1. You cannot say any more about quadrilateral ABCD than you could before.

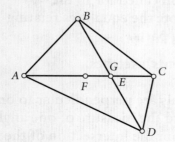

2. When points E, F, and G coincide, ABCD becomes a parallelogram. You cannot say what special type of parallelogram it is. For example, you cannot say for certain that it is a rectangle.

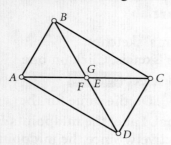

3. Adjust the diagonals so that they are perpendicular to one another and keep E, F, and G coincident. Find and display m∠AFB and make adjustments until it is 90°.

   m∠AFB = 90°

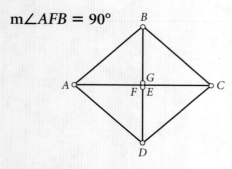

4–6. Check students' sketches.

6. b. Answers may vary. A parallelogram has two pairs of opposite sides congruent. A rectangle has congruent diagonals. A rhombus has perpendicular diagonals.

## Lesson 4.8

1. a. Check students' sketches.
   b. Reflections preserve congruence.
   c. Mark point X on line k. Choose line k as the line of reflection. Reflect $\overline{AX}$ across line k. Then $\overline{AX} \cong \overline{CX}$ and the quadrilateral with have two pairs of adjacent sides congruent. Alternatively, draw a line through A and C. Use this as the line of reflection. Reflect $\overline{AB}$ and $\overline{BC}$ across $\overleftrightarrow{AC}$. The result will be a rhombus because all four sides are congruent.

2. a. Check students' sketches.
   b. 45°
   c. Label the image of point Q as point S. Draw the line through S and Q. SQ is the next line of reflection. Reflect $\overline{PQ}$ and $\overline{PS}$ across $\overleftrightarrow{SQ}$. Label the image of point P as point R.

   m∠QPX = 45.3°

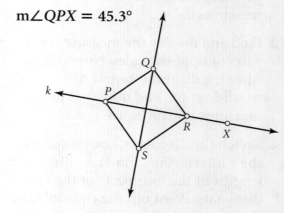

3. $\dfrac{(n-2)180}{2n}$; 60°

4. Check students' sketches.

# Answers

5. Reflections preserve length and angle measure, there will be 6 congruent sides and angles.

6. Rotate $\overline{AB}$ about $O$ through a 60° angle, in the same direction each time. The result will be six points. In the given diagram, these are $A$, $B$, $C$, $D$, $E$, and $F$. Join $A$ to $C$, $C$ to $E$, $E$ to $B$, $B$ to $D$, $D$ to $F$, and $F$ to $A$.

## Student Technology Guide — Chapter 5

### Lesson 5.2

1. $2(.5 \times 2.84 \times 5.12) \approx 14.54$ square units
2. $4.96 - (2.48\sqrt{2})^2 \approx 12.30$ square units
3. $.5 \times 5.12 (5.12 + 10.80) - (2.56\sqrt{2})^2 \approx 27.65$ square units
4. 15.5 units
5. 15.8 units
6. 16.1 units

### Lesson 5.3

1. If $ABCD$ is a rectangle and its vertices are right angles, then $ABCD$ is a rectangle. $ABCD$ is a parallelogram, so $AD = BC$ and $AB = DC$. Since $AB = AD$, then $AB = BC = CD = DA$. Thus, $ABCD$ is a square.

2. Check students' sketches.

3. The ratio of areas does not change when the dynamic sketch changes, but the difference of the areas does change.

4. a. $\dfrac{\text{area circle } E}{\text{area square } ABCD} = \dfrac{\pi\left(\frac{s}{2}\right)^2}{s^2} = \dfrac{\pi}{4}$, which is constant.
   (Area square $ABCD$) − (area circle $E$) = $s^2 - \pi\left(\frac{s}{2}\right)^2$, which is a function of $s$.

   b. From part **a**, the ratio $\frac{\pi}{4}$ is constant. The difference changes as the size of the square changes.

5. a. Answers may vary. The ratio involves quantities related to side length of a square and part of a diagonal, so the ratio is fixed if those quantities divide out.

   b. The difference involves lengths of parts of the circle and square; but $\pi$ is involved, so the difference varies because the variable does not subtract out.

6. Check students' sketches.

7. They show that the ratio of areas is fixed at $\frac{2}{\pi}$. The difference is indeed variable; it equals $(\pi - 2)(AB)^2$.

8. The conjectures made in Exercise 5 are true for these as well. The ratio is equivalent to $\dfrac{\sqrt{2}}{\pi}$, and the difference is $(2\pi - 4\sqrt{2})AB$.

### Lesson 5.4

1. 132.3 units
2. 1104.7 units
3. 10.7 units
4. 1.8 units
5. 11.1 units

# Answers

6. a. When the expression for $x + y$ is changed to $x - y$, the display shows a negative number. Since $x - y < 0, x < y$.
   b. $\approx 0.94$ units, which is the absolute value of $x - y$.

7. a. $\sqrt{(4.5^2 + 4.5^2)} + \sqrt{(4.5^2 + 4.5^2)}$, or $2(4.5)\sqrt{2}$
   b. about 12.73 units

## Lesson 5.5

1. 259.8 square units

2. 1496.5 square units

3. 3367.1 square units

4. area $= \dfrac{3r^2\sqrt{3}}{2}$

5. 259.8 square units

6. 1496.5 square units

7. 3367.1 square units

8. a. area $= \left[2\left(\dfrac{50}{\sqrt{2}}\right)\right]^2$
   b. 5000 square units

9. area $= (r\sqrt{2})^2$; 5000 square units

## Lesson 5.6

1. Check students' spreadsheets.

2. The entry is greater than $\pi$ because the rectangles extend beyond the circular region's boundary. The approximation is roughly 0.198 more than $\pi$.

3. In cell B2, enter =SQRT(1-((A2-1)/100)^2)/100. Fill down to row 101. Change the entry in cell C2 to =4*SUM(B2:B101).
   Total area $\approx 3.16042$.

## Lesson 5.7

1. Check students' sketches. Refer to the diagram below for the answers to Exercises 2–5.

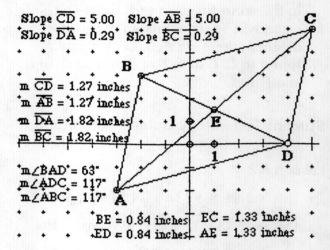

2. The diagram above shows that the slopes of $\overline{CD}$ and $\overline{BA}$ are equal and that the slopes of $\overline{AD}$ and $\overline{BC}$ are equal. Thus, by definition, ABCD is a parallelogram.

3. The diagram above shows that the lengths of $\overline{CD}$ and $\overline{BA}$ are equal and that the lengths of $\overline{AD}$ and $\overline{BC}$ are equal. Opposite sides of a quadrilateral are equal in length, so ABCD is a parallelogram.

4. $\angle BAD$ and $\angle ADC$ are supplementary, so $\overline{AB} \parallel \overline{CD}$. $\angle BAD$ and $\angle ADC$ are supplementary, so $\overline{AD} \parallel \overline{BC}$. Opposite sides of ABCD are parallel so ABCD is a parallelogram.

5. Sketch diagonals $\overline{AC}$ and $\overline{BD}$. Locate and mark point E, where they intersect. Find and display AE, EC, BE, and ED. Diagonals AC and BD bisect one another, so ABCD is a parallelogram.

6. Using any of the strategies in Exercises 2–5, WXYZ is a parallelogram.

# Answers

7. Check students' sketches.

8. The sketch below shows that △XYZ is a scalene triangle. The measurement of ∠YXZ indicates that △XYZ is almost a right triangle.

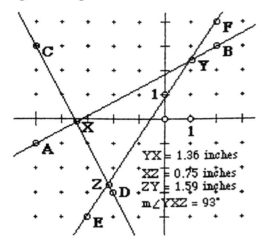

### Lesson 5.8

1. $\frac{\pi}{4} \approx 0.785398163$

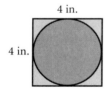

2. $\frac{\pi}{6} \approx 0.523598776$

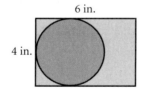

3. $\frac{\pi}{8} \approx 0.392699082$

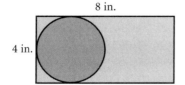

4. As $a$ increases, the maximum probability decreases.

5. Circle: 6.536 square inches; triangle: 10.851 square inches; the maximum probability is 0.602340798. The maximum probability occurs when the circle is the inscribed circle of the triangle.

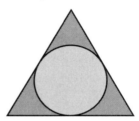

6. It is the inscribed circle of a regular polygon which maximizes the probability a point chosen at random inside the polygon is also inside the circle. (Check students' sketches of the regular pentagon and inscribed circle. The maximum probability is 0.8655408331.)

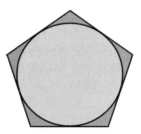

## Student Technology Guide — Chapter 6

### Lesson 6.3

1. Check students' sketches.

2. Translations preserve congruence. Thus, the image of the front is congruent to the front. Since each vertex in square $ABCD$ is translated by the same amount in the same direction, each segment joining back to front is congruent to all such segments.

# Answers

3. Check students' sketches.

4. For the triangular prism, drag point A directly right and point D directly left until they coincide. This turns the quadrilateral into a triangle. The software carries out the same dragging in the back face of the prism.
   For the trapezoidal prism, drag only point D directly to the left by a small amount, keeping it directly to the right of point A. The software carries out the same dragging in the back face of the prism.

5. Check students' models.

6. Using $\overline{BX}$ as an axis of rotation, rotate the box clockwise 90° so that DBXZ lies on the desk. Using $\overline{DZ}$ as an axis of rotation, rotate clockwise 90°. The result will be that CDZY lies on the desk and point B is at the top, in the front, and to the left.

7. Check students' sketches.

8. Actions and choices may vary. Using $\overline{BX}$ as an axis of rotation, rotate the box counterclockwise 90° so that face ACYW lies on the desk. Using $\overline{CY}$ as an axis of rotation, rotate clockwise 90°. The result will be that CDZY lies on the desk and point B is at the top, in the front, and to the left. The computer display that corresponds to this set of rotations is shown below.

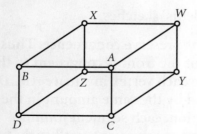

## Lesson 6.6

1. Check students' sketches.

2. In the sketch at the left in the display, the viewer is in front of the box and to its right. In the sketch at the right in the display, the viewer is in front of the box but to its left.

3. Drag point P to the center (intersection of the diagonals) of the front face. The solid border is visible and the dashed border is not.

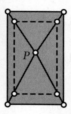

4. Locations and descriptions may vary. The back becomes the front. In the sketch below, the the box is between you and the viewer and the viewer is to your left.

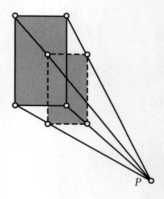

# Answers

The back becomes the front. In the sketch below, the box is between you and the viewer and the viewer is to your right.

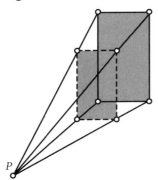

5. Check students' sketches.

6. Find and display the midpoint of $\overleftrightarrow{PQ}$. Draw the line perpendicular to $\overleftrightarrow{PQ}$ at the midpoint. Drag point P directly right or point Q directly left until the line perpendicular to $\overleftrightarrow{PQ}$ contains $\overline{AB}$.

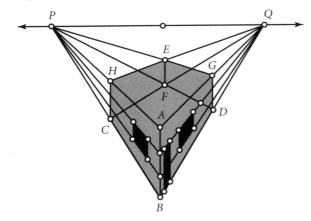

7. Sketches and descriptions may vary. A one-point perspective of the house is shown below.

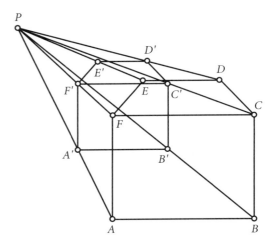

It was drawn by choosing a vanishing point P, then drawing $\overline{AP}, \overline{BP}, \overline{CP}, \overline{DP}, \overline{EP}$, and $\overline{FP}$. The back of the lower portion of the house was drawn using the directions in Exercise 1. To draw the back of the trapezoidal prism for the top, draw a line through F' parallel to $\overline{EF}$. Mark the intersection of that line with $\overline{EP}$ as E'. Then complete the sketch by drawing $\overline{E'D'}, \overline{D'C'}$, and $\overline{C'F'}$.

**Student Technology Guide — Chapter 7**

**Lesson 7.3**

1. a. $V_b = x \times x \times \frac{1}{4}x = \frac{1}{4}x^3$

   b. $V_p = \frac{1}{3}\left[(10h)\left(\frac{1}{2}x\right)\left(\frac{1}{2}x\right)\right] = \frac{1}{3}\left[\left(\frac{10}{4}x\right)\left(\frac{1}{2}x\right)\left(\frac{1}{2}x\right)\right] = \frac{10}{48}x^3$

   c. $V = V_b + V_p = \frac{1}{4}x^3 + \frac{10}{48}x^3 = \frac{22}{48}x^3$

2. a. 458.33 cubic feet
   b. 792 cubic feet
   c. 2673 cubic feet
   d. 3666 cubic feet

# Answers

3. In cubic yards, $V = \frac{1}{27}\left(\frac{22}{48}\right)x^3 = \frac{22}{48 \times 27}x^3$; 63.21 cubic yards

4. a. 426.25 square feet
   b. 613.80 square feet
   c. 1381.04 square feet
   d. 1704.99 square feet

5. The expression $(0.1x)(x)(0.5x)$ represents the volume of the bottom slab, so $2[(0.1x)(x)(0.5x)]$ represents the volume of both slabs. The volume of one column is $(0.1x)^2(1.5)$, so $4[(0.1x)^2(1.5)]$ represents the volume of the four columns.

6. a. 34.56 cubic feet
   b. 54.88 cubic feet
   c. 81.92 cubic feet
   d. 116.64 cubic feet

7. a. 8.8 feet
   b. 9.1 feet
   c. 9.3 feet
   d. 9.6 feet

## Lesson 7.4

1. Check students' spreadsheets.

2. a. $S = 2\pi(2hr - ht + 2rt - t^2)$
      $V = \pi h(2rt - t^2)$
   b. $= 2*3.14159*(2*D2*A2-2*D2*B2-2*A2*B2-B2\wedge 2)$
      $= 3.14159*D2*(2*A2*B2-B2\wedge 2)$

3. From the spreadsheet, surface area increases and volume decreases as thickness decreases.

| E | F |
|---|---|
| S | V |
| 4380.38177 | 437.31 |
| 4387.5691 | 416.16 |
| 4394.75517 | 394.94 |
| 4401.93999 | 373.64 |
| 4409.12355 | 352.26 |
| 4416.30585 | 330.81 |
| 4423.4869 | 309.26 |
| 4430.66669 | 287.68 |
| 4437.84522 | 266.00 |
| 4445.0225 | 244.25 |
| 4452.19852 | 222.42 |

4. $S = 2\pi(r^2 + 2hr - ht)$, or $\pi(2r^2 - s^2) + 2\pi(rh - sh + sh')$
   $V = \pi h(2rt - t^2) + \pi h'(r^2 - 2rt + t^2)$, or $\pi(r^2h - s^2h + s^2h')$

5. Again, surface area increases and volume decreases as thickness decreases.

| E | F |
|---|---|
| S | V |
| 4482.67 | 6338.67 |
| 4490.38 | 6360.25 |
| 4498.08 | 6381.90 |
| 4505.79 | 6403.64 |
| 4513.49 | 6425.47 |
| 4521.20 | 6447.37 |
| 4528.91 | 6469.36 |
| 4536.62 | 6491.42 |
| 4544.34 | 6513.58 |
| 4552.05 | 6535.81 |
| 4559.77 | 6558.13 |

## Lesson 7.5

1. 2094.4 cubic units

2. 23,090.7 cubic units

3. 14,476.5 in.$^3$

4. 104.7 cubic feet

5. 1055.6 in.$^3$

6. 1205.2 in.$^3$

7. 848.9 in.$^3$

8. 1206.2 in.$^3$

# Answers

9. 1205.2 in.$^3$

10. 326.9 in.$^3$

**Lesson 7.6**

1. 226.2 in.$^2$; 904.8 in.$^3$
2. 106.8 in.$^2$; 53.4 in.$^3$
3. 226.2 in.$^2$; 150.8 in.$^3$
4. 238.4 in.$^2$; 155.6 in.$^3$
5. 603.2 in.$^2$; 2412.7 in.$^3$
6. 904.8 in.$^2$; 1809.6 in.$^3$

**Student Technology Guide — Chapter 8**

**Lesson 8.1**

1. Check students' sketches.

2. Rotate $\triangle ABC$ about $O$ clockwise 180°. Then, using $O$ as a center of dilation, enlarge the image by a scale factor of 1.5.

3. As long as the scale factor is $-1.5$, the observations from Exercise 2 stay the same.

4. $s > 1$: The figure is enlarged and there is no rotation.
$s = 1$: The dilation has no effect on the original figure.
$0 < s < 1$: The figure is shrunk or contracted and there is no rotation.
$s = 0$: The image is a point, the center of the dilation.
$-1 < s < 0$: The figure is shrunk and rotated 180°.
$s = -1$: The figure is rotated 180° and there is no enlargement or contraction.
$s < -1$: The figure is rotated 180° and is enlarged.

5. Check students' sketches.

6. Point $A$ and all of its images lie along the line $y = \frac{1}{2}x$ and each successive image gets closer to the origin. The origin is the limit point of the set of successive images.

7. As in Exercise 6, point $A$ and all of its images lie along the line $y = \frac{1}{2}x$. However, the successive images get farther and farther from the origin rather than closer.

8. To get point $A$ and its successive images to bounce back and forth between Quadrants 1 and III, Debra needs a negative scale factor of $s$. Based on Exercises 6 and 7, Debra needs to choose a value of $s$ such that $s \leq -1$. If she chooses $-1 < s < 0$, the successive images will get closer and closer to one another rather than farther and farther apart. In the diagram below, $s = -1.25$.

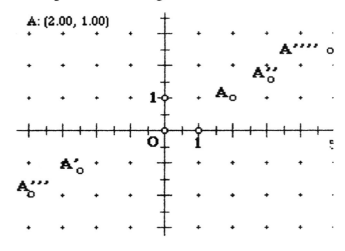

**Lesson 8.2**

1. a. 2.05
   b. $NO = 1.22$, $OP = 1.70$, $PQ = 3.71$, $QR = 1.59$

2. $XY = 3.09$, $YZ = 3.48$, $ZW = 3.04$

# Answers

3. $PQ = 3.71$, $QR = 1.98$, $RS = 2.45$, $ST = 2.45$

4. $YZ = 108$, $XZ = 81$

## Lesson 8.4

1. Check students' sketches.

2. Explorations may vary. Find and display the lengths of all sides of $\triangle ABC$ and $\triangle ADE$. By the SSS Triangle Similarity Theorem, $\triangle ABC \sim \triangle ADE$.

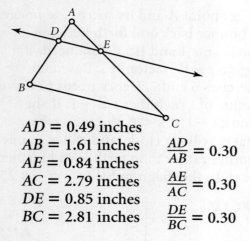

$AD = 0.49$ inches
$AB = 1.61$ inches
$AE = 0.84$ inches
$AC = 2.79$ inches
$DE = 0.85$ inches
$BC = 2.81$ inches

$\frac{AD}{AB} = 0.30$
$\frac{AE}{AC} = 0.30$
$\frac{DE}{BC} = 0.30$

3. The claim is not stated with sufficient precision. The claim must also state in the hypothesis that the line parallel to one side of the triangle intersects each of the other sides at distinct points. It must rule out a line that does not intersect the triangle at all and a line that intersects the triangle at a vertex.

4. a. Check students' sketches.
   b. The diagram that results is essentially the same as the diagram sketched in Exercise 1, with $s > 1$. If two corresponding angles are congruent in $\triangle ABC$ and $\triangle ADE$, then $\triangle ABC \sim \triangle ADE$ by the AA Triangle Similarity Postulate.

## Lesson 8.5

1. Check students' sketches.

2. Check students' calculations.

3. Ratios relating the scale diagram and the actual physical layout on the land was used together with the scale 1 inch:100 meters.

4. After making the adjustments, find and display $AC$ and $CF$. Apply the scale. The actual distances are 144 meters and 100 meters respectively.

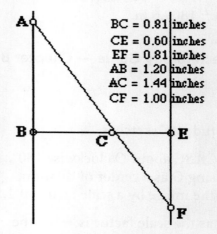

5. Let 1 inch represent 100 units. The scale drawing is shown below.

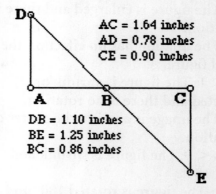

From the drawing, $DB = 110$ units, $BE = 125$ units, and $BC = 86$ units.

# Answers

6. a. Scales may vary. Sample scale: 1 inch represents 2 feet.
   b. Check students' sketches.
   c. Check students' sketches.

7–13. Answers may vary.

7. 3.5 inches

8. 1.875 inches, about 1.88 inches

9. 3 inches

10. 1 inch

11. 3 inches long and 1.25 inches deep

12. 1.5 inches

13. a. 38.34 square inches
    b. 1 square inch represents 4 square feet.
    c. about 153.38 square feet

## Lesson 8.6

1. The flow rate through $P_1$ is 16 times that through $P_2$.

2. The flow rate through $P_1$ is about one sixteenth times that through $P_2$.

3. $F_1$ to $F_2$: about 5; $F_2$ to $F_3$: about 5; $F_1$ to $F_3$: about 26

4. The radius of $P_1$ is twice that of $P_2$.

## Student Technology Guide — Chapter 9

### Lesson 9.1

1. Check students' sketches.

2. a. Adjust the points so that $m\widehat{AB} = m\widehat{BC} = m\widehat{CA} = 120°$. Then the arcs will be congruent. These congruent arcs give $\overline{AB} \cong \overline{BC} \cong \overline{CA}$.

   b. This congruence statement says that $\triangle ABC$ is equilateral.

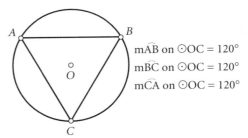

$m\widehat{AB}$ on $\odot OC = 120°$
$m\widehat{BC}$ on $\odot OC = 120°$
$m\widehat{CA}$ on $\odot OC = 120°$

3. a. Adjust the points on the circle so that each arc has measure $\frac{360°}{5} = 72°$. Then $m\widehat{HI} = m\widehat{IJ} = m\widehat{JK} = m\widehat{KL} = m\widehat{LH} = 72°$, so $\overline{HI} \cong \overline{IJ} \cong \overline{JK} \cong \overline{KL} \cong \overline{LH}$.

   b. The five triangles formed with point $P$ as common vertex are congruent by SSS. Then by the Angle Addition Postulate, the interior angles of the pentagon are congruent; so HIJKL is a regular pentagon.

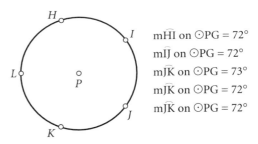

$m\widehat{HI}$ on $\odot PG = 72°$
$m\widehat{IJ}$ on $\odot PG = 72°$
$m\widehat{JK}$ on $\odot PG = 73°$
$m\widehat{JK}$ on $\odot PG = 72°$
$m\widehat{JK}$ on $\odot PG = 72°$

4. The strategy described is the generalization of the method used in Exercises 2 and 3. If the arcs have measures of $\frac{360°}{n}$, the chords determined by the $n$ points will be congruent. By reasoning similar to that in part **b** of Exercise 3, the interior angles will be congruent. Thus, the polygon will be a regular polygon.

# Answers

**5. a.** The entry in cell B17 indicates that after 75 seconds, each point has traveled 375°, or more than one full revolution.

**b.** Although the angle through which each point travels is the same, the distance traveled increases as the radius increases.

**6.** $\frac{R_2}{R_1} = \frac{D_2}{D_1} = 2$ and $\frac{R_3}{R_1} = \frac{D_3}{D_1} = 3$. The distance traveled is proportional to the radius of the circle on which points travel.

**7.** Check students' spreadsheets. Yes; if $R_2$ is twice $R_1$, and $R_3$ is three times $R_1$, then $D_2 = 2D_1$ and $D_3 = 3D_1$.

**8.** If $s = 1$, then the distances are the same. If $s > 1$, then the point on $C_2$ travels farther. If $0 < s < 1$, then the point on $C_1$ travels farther.

## Lesson 9.2

**1.** Check students' sketches.

**2.** Equilateral; Measuring the side lengths of $\triangle ABC$ indicates that $AB = BC = CA$.

**3. a.** $\triangle ABC \sim \triangle GHF$.
**b.** Find and display the lengths of the sides of $\triangle ABC$ and $\triangle GHF$; $\frac{AB}{HF} = \frac{AC}{GH} = \frac{BC}{GF}$. By the SSS Triangle Similarity Theorem, $\triangle ABC \sim \triangle GHF$.

**4.** If the process is continued indefinitely, it generates a sequence of equilateral triangles. All of them are similar to $\triangle ABC$ and similar to one another.

**5.** Check students' sketches.

**6.** square

**7. a.** Squares $ABCD$ and $GHIF$ are similar.
**b.** Find and display the side lengths of the two quadrilaterals. Find the ratios of corresponding sides. The ratios will be equal. Find and display the measures of the corresponding angles. They are congruent.

**8.** Successive constructions would produce a sequence of squares. All squares would be similar to one another.

**9.** If $n$ points are scattered around the circle and are adjusted so that the central angles are equal, then the sequence of $n$-gons produced would be similar.

## Lesson 9.3

**1.** Check students' sketches. If two inscribed angles intercept the same arc, the angles are congruent and the measure of the intercepted arc is twice that of the inscribed angles.

**2.** $\frac{AB}{DC} = \frac{AE}{DE} = \frac{BE}{CE}$; by the SSS Triangle Similarity Theorem, $\triangle ABE \sim \triangle DCE$.

**3.** Display m$\angle ABE$ and m$\angle ACD$. They intercept the same arc, so they have the same measure. This fact, together with m$\angle AEB$ = m$\angle DEC$, is sufficient to conclude that $\triangle ABE \sim \triangle DCE$ by the AA Triangle Similarity Postulate.

**4.** Yes. The reasoning in Exercises 2 and 3 applies here too. Since $\angle AED$ and $\angle BEC$ are vertical angles, their measures are equal. Measure $\angle DAC$ and $\angle CBD$. They will have the same measure. By the AA Triangle Similarity Postulate, $\triangle AED \sim \triangle BEC$.

# Answers

5. In the diagram below, each of the inscribed angles is a right angle. Thus, A, B, C, and D form a rectangle inscribed in the circle.

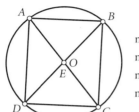

m∠DAB = 90°
m∠ABC = 90°
m∠BCD = 90°
m∠CDA = 90°

6. The sum is 720°. Since the sum of the related inscribed angles is one-half this sum, the sum of the measures of the interior angles of a convex quadrilateral is 360°.

## Lesson 9.4

1. Check students' sketches.

2. a. Check students' sketches.
   b. The sum of the measures of the five points in a five-pointed star is 180°.

3. $m\angle A = \frac{1}{2}\left[m\widehat{WX} + m\widehat{XY} + m\widehat{YZ} - \left(m\widehat{ZV} + m\widehat{VW}\right)\right]$

   $m\angle B = \frac{1}{2}\left[m\widehat{XY} + m\widehat{YZ} + m\widehat{ZV} - \left(m\widehat{VW} + m\widehat{WX}\right)\right]$

   $m\angle C = \frac{1}{2}\left[m\widehat{YZ} + m\widehat{ZV} + m\widehat{VW} - \left(m\widehat{WX} + m\widehat{XY}\right)\right]$

   $m\angle D = \frac{1}{2}\left[m\widehat{ZV} + m\widehat{VW} + m\widehat{WX} - \left(m\widehat{XY} + m\widehat{YZ}\right)\right]$

   $m\angle E = \frac{1}{2}\left[m\widehat{VW} + m\widehat{WX} + m\widehat{XY} - \left(m\widehat{YZ} + m\widehat{ZV}\right)\right]$

   $m\angle A + m\angle B + m\angle C + m\angle D + m\angle E = \frac{1}{2}\left(m\widehat{WX} + m\widehat{XY} + m\widehat{YZ} + m\widehat{ZV} + m\widehat{VW}\right) = \frac{1}{2}(360°) = 180°$

4. The sum of the five outside angles (∠AUB, ∠BQC, ∠CRD, ∠DSE, and ∠ETA) is 540°.

5. The sum of the measures of the angles at the points of a six-pointed star is 180°. If the star has $n$ points, the sum is still 180°. The sum of the $n$ outside angles is 180°($n - 2$).

## Lesson 9.5

1. 5.4    2. 6.2    3. 3.7    4. 13.7

5. $\sqrt{10(2 \times 24 - 10)}$; 19.5

6. $\sqrt{8.5(2 \times 30 - 8.5)}$; 20.9

7. $\sqrt{100(2 \times 100 - 100)}$; 100

## Lesson 9.6

1. $y = \pm\sqrt{16 - x^2}$

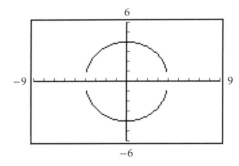

2. $y = \pm\sqrt{36 - x^2}$

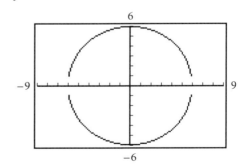

# Answers

3. $y = \pm\sqrt{49 - x^2}$

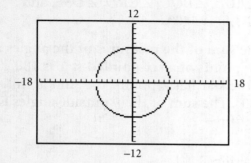

4. $y = \pm\sqrt{64 - x^2}$

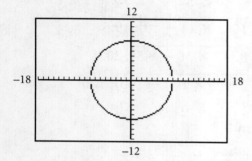

5. $y = \pm\sqrt{81 - x^2}$

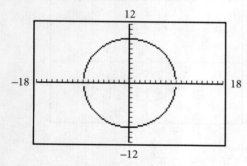

6. $y = \pm\sqrt{100 - x^2}$

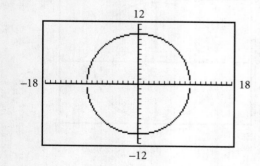

7. $y = 2 \pm \sqrt{9 - (x - 1)^2}$

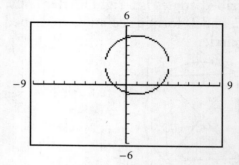

8. $y = -2 \pm \sqrt{9 - (x + 1)^2}$

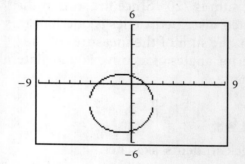

9. $y = -2 \pm \sqrt{9 - (x - 1)^2}$

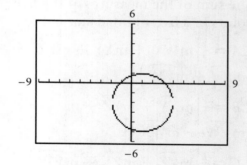

10. a.

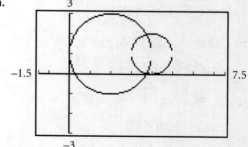

b. two points
c. approximate coordinates: (3.74, 1.98) and (3.74, 0.02)

# Answers

11. a.–b.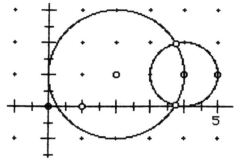

c. two points

12. a. large circle: radius of 4 and center at $(0, -1)$
small circle: radius of 3 and center at $(-1, -2)$

b.–c. $(-3.96, -1.54); (-0.54, -4.96)$

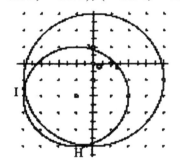

## Student Technology Guide—Chapter 10

### Lesson 10.1

1. 1
2. 3.1910
3. 0.4122
4. 1.4715
5. 5.6713
6. 0.1763
7. 0.5774
8. 1.7321

9. 7.2°
10. 41.4°
11. 53.6°
12. 6.3°
13. 45°
14. 22.5°
15. 50.8°
16. 7.8°
17. The statement is true. The value of the tangent increases as the angle measure increases. As the tangent ranges from 0 to 1, the measure of the angle ranges from 0° to 45°.

### Lesson 10.2

1. $KM \approx 5.3; LM \approx 4.7$
2. $YW \approx 2.9; XW \approx 4.2; WZ \approx 6.2$
3. $PS \approx 4.4; QR \approx 4.4$
4. 37.8°
5. 23.1°
6. 52.2°
7. 142.2°
8. 66.9°
9. 14.7°
10. Answers will vary. [ 2nd ] [ COS ] 5.36 [ ÷ ] 6.91 [ ) ] [ENTER] gives 39.1°.

### Lesson 10.4

1. one
2. two
3. one

# Answers

**4.** none

**5.** m∠Z = 70°, m∠Y = 40°, XZ ≈ 6.8

**6.** m∠X = 20°, m∠Y = 140°, XZ ≈ 37.6

**7.** m∠X = 55°, m∠Z = 55°, XZ ≈ 28.7

**8.** No; the equation $\frac{\sin 80°}{12} = \frac{\sin M}{36}$ implies that sin M = 3 sin 80° ≈ 2.9. Since sin M ≤ 1, no such triangle exists.

## Lesson 10.5

**1.** m∠A ≈ 41.4°; m∠B ≈ 41.4°; m∠C ≈ 97.2°

**2.** m∠A ≈ 20.7°; m∠B ≈ 32.1°; m∠C ≈ 127.2°

**3.** m∠A ≈ 39.6°; m∠B ≈ 49.9°; m∠C ≈ 90.5°

**4.** m∠A ≈ 75.5°; m∠B ≈ 75.5°; m∠C ≈ 29.0°

**5.** m∠A ≈ 41.6°; 59.8 square units

**6.** m∠A ≈ 15.4°; 27.8 square units

## Lesson 10.6

**1.** Check students' sketches.

**2.**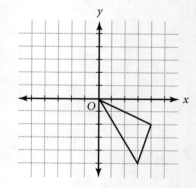
OP = 2.00 inches
OQ = 2.00 inches
m∠POZ = 25°
m∠QOZ = 65°
OR = 3.76 inches
m∠ROZ = 45°

**3.**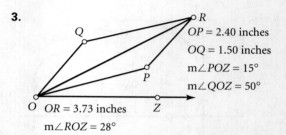
OP = 2.40 inches
OQ = 1.50 inches
m∠POZ = 15°
m∠QOZ = 50°
OR = 3.73 inches
m∠ROZ = 28°

**4.** Check students' sketches.

**5.** (9, 4)

**6.** (5, 5)

**7.** (3, 3)

**8.** (2, −2)

**9.** (0, 0)

**10.** (0, 0)

**11. a.** $(a + c, b + d)$
   **b.** (14.9, 6.8)
   **c.** If one vector has the coordinates $(a, b)$, then the other has the coordinates $(-a, -b)$.

## Lesson 10.7

**1.** $\begin{bmatrix} 0 & 4 & 3 \\ 0 & -2 & -5 \end{bmatrix}$

# Answers

2. $\begin{bmatrix} 0 & 4 & 3 \\ 0 & -2 & -5 \end{bmatrix}$

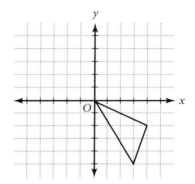

3. Since $(R_{45})^8 = \begin{bmatrix} 1 & 0 \\ 0 & 1 \end{bmatrix}$, $(R_{45})^{(8n)}B = B$.

## Student Technology Guide—Chapter 11

### Lesson 11.3

1. a. 2; The entry in row 2 and column 5 of $A^2$ is also 2.
   b. $A^2$ indicates how many communication paths there are from one member to another via one intermediary.

2. a. $A^3 = \begin{bmatrix} 3 & 4 & 4 & 0 & 4 \\ 4 & 3 & 2 & 0 & 2 \\ 4 & 2 & 3 & 0 & 2 \\ 5 & 4 & 4 & 0 & 3 \\ 4 & 2 & 2 & 0 & 3 \end{bmatrix}$
   b. 4; the entry in row 3 and column 1 of $A^2$ is also 4.
   c. $A^3$ indicates how many communication paths there are from one member to another via two intermediaries.

### Lesson 11.4

1. The line segment stays the same. It does not bend.

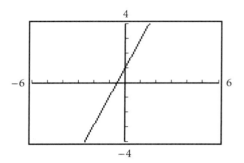

2. The line segment becomes V shaped.

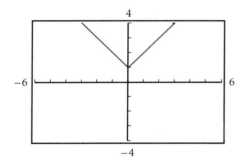

3. The line segment becomes U shaped.

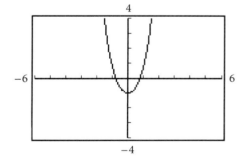

# Answers

4. The line segment becomes S shaped.

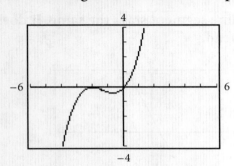

5. The line segment becomes a semicircle.

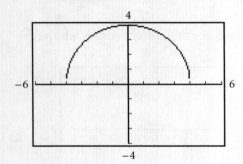

6. The line segment becomes W shaped.

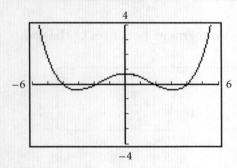

7. The interval becomes splits into three branches.

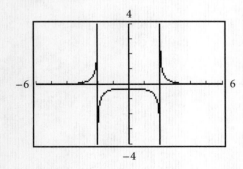

8. a. Check students' sketches.
   b.
   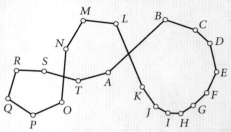
   c. In the original figure, you can get from any interior point to any other interior point without crossing the border to do it. In the second figure, this is not possible. Choose a point in one loop and a point inside the other loop. Any line segment joining them must cross a side of the figure.

9. a. Check students' sketches.
   b. Drag vertices so that exactly three consecutive vertices are collinear. In effect, move one vertex so that it lies along the line containing the vertex before it and the vertex after it in the sequence A, B, C, D, E, F, G, H, I, J, K, L, M, N, O, P. Both the original figure and the new one have only one interior part.

   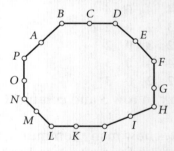

   c. Both the original figure and the new one have only one interior part.

   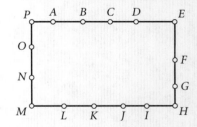

# Answers

10. A dodecagon has 12 vertices, so start with 12 points arranged around a rectangle. Four of the points will be vertices and the other points will be along the sides. Then drag points so that 8 points become vertices and the others lie along sides. Then drag points so that all 12 points become vertices.

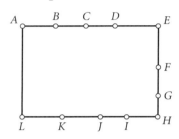

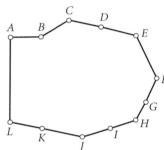

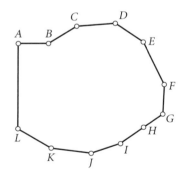

### Lesson 11.6

1. $s = 3^{n-1}$
2. $L = \left(\frac{1}{2}\right)^{n-1}$
3. $P = 3Ls = 3\left(\frac{3}{2}\right)^{n-1}$

4. Check students' spreadsheets.

| n | s | L | P |
|---|---|---|---|
| 1 | 1 | 1 | 3 |
| 2 | 3 | 0.5 | 4.5 |
| 3 | 9 | 0.25 | 6.75 |
| 4 | 27 | 0.125 | 10.125 |
| 5 | 81 | 0.0625 | 15.1875 |
| 6 | 243 | 0.03125 | 22.7813 |
| 7 | 729 | 0.01563 | 34.1719 |
| 8 | 2187 | 0.00781 | 51.2578 |
| 9 | 6561 | 0.00391 | 76.8867 |
| 10 | 19683 | 0.00195 | 115.33 |

5. $P$ increases without bound as $n$ increases.

6. a. Check students' spreadsheets. The total area is given by the expression below.

$$A = (0.43301)(3^{n-1})\left[\left(\frac{1}{2}\right)^{n-1}\right]^2$$

| n | s | L | A |
|---|---|---|---|
| 1 | 1 | 1 | 0.43301 |
| 2 | 3 | 0.5 | 0.32476 |
| 3 | 9 | 0.25 | 0.24357 |
| 4 | 27 | 0.125 | 0.18268 |
| 5 | 81 | 0.0625 | 0.13701 |
| 6 | 243 | 0.03125 | 0.10276 |
| 7 | 729 | 0.01563 | 0.07707 |
| 8 | 2187 | 0.00781 | 0.0578 |
| 9 | 6561 | 0.00391 | 0.04335 |
| 10 | 19683 | 0.00195 | 0.03251 |

b. As $n$ increases, the total area of the shaded triangles decreases and approaches 0.

# Answers

**7. a.** Check students' spreadsheets.

| n | s | L | A |
|---|---|---|---|
| 1 | 1 | 1 | 0 |
| 2 | 3 | 0.5 | 0.10825 |
| 3 | 9 | 0.25 | 0.18944 |
| 4 | 27 | 0.125 | 0.25033 |
| 5 | 81 | 0.0625 | 0.296 |
| 6 | 243 | 0.03125 | 0.33025 |
| 7 | 729 | 0.01563 | 0.35594 |
| 8 | 2187 | 0.00781 | 0.37521 |
| 9 | 6561 | 0.00391 | 0.38966 |
| 10 | 19683 | 0.00195 | 0.4005 |

**b.** As $n$ increases without bound, the total area of the unshaded region approaches the area of the original triangle, 0.43301 square units.

**8.** $S_5 = 1 - \left(\frac{1}{3}\right)^2 - 8\left[\left(\frac{1}{3}\right)\left(\frac{1}{3}\right)\right]^2 - 64\left[\left(\frac{1}{3}\right)\left(\frac{1}{3}\right)\left(\frac{1}{3}\right)\right]^2 - 512\left[\left(\frac{1}{3}\right)\left(\frac{1}{3}\right)\left(\frac{1}{3}\right)\left(\frac{1}{3}\right)\right]^2$

**9. a.** From $S_n$, subtract the next third power of 2 times the square of the product of $\frac{1}{3}$ with itself one more time than in $S_{n-1}$.
**b.** $S_n = S_{n-1} - (2^{3(n-2)})\left(\frac{1}{3}\right)^{2(n-1)}$

**10. a.** Check students' spreadsheets.

**b.**

| n | Sn |
|---|---|
| 1 | 1 |
| 2 | 0.888888889 |
| 3 | 0.790123457 |
| 4 | 0.702331962 |
| 5 | 0.624295077 |
| 6 | 0.554928957 |
| 7 | 0.493270184 |
| 8 | 0.438462386 |
| 9 | 0.389744343 |
| 10 | 0.346439416 |
| 11 | 0.307946148 |
| 12 | 0.273729909 |
| 13 | 0.243315475 |
| 14 | 0.216280422 |

**c.** $S_n$ decreases and approaches 0 as $n$ increases.

## Lesson 11.7

**1. a.** Interchange 1.5 and 2.

**b.** $\begin{bmatrix} 2 & 0 \\ 0 & 1.5 \end{bmatrix}$

**c.** $AB = \begin{bmatrix} 0 & 0 & 8 & 8 \\ 0 & 6 & 6 & 0 \end{bmatrix}$; the vertices are $O(0, 0)$, $A'(0, 6)$, $B'(8, 6)$, and $C'(8, 0)$; the horizontal extent is 8 units and the vertical extent is 6 units.

# Answers

2–3. $CB = \begin{bmatrix} 0 & 8 & 12 & 4 \\ 0 & 4 & 4 & 0 \end{bmatrix}$;

$DB = \begin{bmatrix} 0 & 0 & 4 & 4 \\ 0 & 4 & 12 & 8 \end{bmatrix}$;

Matrix $CB$ skews the square into a parallelogram by pushing $\overline{AB}$ to the right but keeping $\overline{OC}$ where it is; matrix $DB$ skews the square into a parallelogram by pushing $\overline{BC}$ up and keeping $\overline{OA}$ where it is.

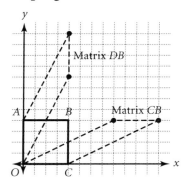

4. $EB = \begin{bmatrix} 0 & -6 & 2 & 8 \\ 0 & 4 & 4 & 0 \end{bmatrix}$; the image is a parallelogram with vertices $A$ and $B$ pushed to the left and vertex $C$ pushed to the right.

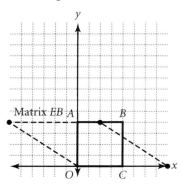

## Student Technology Guide — Chapter 12

### Lesson 12.2

1.
| P | Q | EXP 1 | EXP 1 |
|---|---|---|---|
| 1 | 1 | 0 | 0 |
| 1 | 0 | 0 | 0 |
| 0 | 1 | 0 | 0 |
| 0 | 0 | 1 | 1 |

2.
| P | Q | EXP 1 | EXP 1 |
|---|---|---|---|
| 1 | 1 | 1 | 1 |
| 1 | 0 | 0 | 0 |
| 0 | 1 | 0 | 0 |
| 0 | 0 | 0 | 0 |

3.
| P | Q | EXP 1 | EXP 1 |
|---|---|---|---|
| 1 | 1 | 1 | 1 |
| 1 | 0 | 1 | 1 |
| 0 | 1 | 1 | 1 |
| 0 | 0 | 0 | 0 |

4.
| P | Q | EXP 1 | EXP 1 |
|---|---|---|---|
| 1 | 1 | 1 | 1 |
| 1 | 0 | 1 | 1 |
| 0 | 1 | 0 | 1 |
| 0 | 0 | 0 | 0 |

5. $\sim(p \text{ OR } q)$

| P | Q | EXP 1 | EXP 1 |
|---|---|---|---|
| 1 | 1 | 0 | 0 |
| 1 | 0 | 0 | 0 |
| 0 | 1 | 0 | 0 |
| 0 | 0 | 1 | 1 |

6. Check students' spreadsheets.

# Answers

7.

| P | Q | R | EXP 1 | EXP 2 |
|---|---|---|-------|-------|
| 1 | 1 | 1 | 1 | 1 |
| 1 | 1 | 0 | 1 | 1 |
| 1 | 0 | 1 | 1 | 1 |
| 1 | 0 | 0 | 0 | 0 |
| 0 | 1 | 1 | 0 | 0 |
| 0 | 1 | 0 | 0 | 0 |
| 0 | 0 | 1 | 0 | 0 |
| 0 | 0 | 0 | 0 | 0 |

equivalent

8.

| P | Q | R | EXP 1 | EXP 2 |
|---|---|---|-------|-------|
| 1 | 1 | 1 | 1 | 1 |
| 1 | 1 | 0 | 1 | 1 |
| 1 | 0 | 1 | 1 | 1 |
| 1 | 0 | 0 | 1 | 1 |
| 0 | 1 | 1 | 1 | 1 |
| 0 | 1 | 0 | 0 | 0 |
| 0 | 0 | 1 | 0 | 0 |
| 0 | 0 | 0 | 0 | 0 |

equivalent

9.

| P | Q | R | EXP 1 | EXP 2 |
|---|---|---|-------|-------|
| 1 | 1 | 1 | 1 | 1 |
| 1 | 1 | 0 | 1 | 1 |
| 1 | 0 | 1 | 1 | 1 |
| 1 | 0 | 0 | 1 | 1 |
| 0 | 1 | 1 | 1 | 1 |
| 0 | 1 | 0 | 1 | 1 |
| 0 | 0 | 1 | 1 | 1 |
| 0 | 0 | 0 | 0 | 0 |

equivalent

10.

| P | Q | R | EXP 1 | EXP 2 |
|---|---|---|-------|-------|
| 1 | 1 | 1 | 0 | 0 |
| 1 | 1 | 0 | 0 | 0 |
| 1 | 0 | 1 | 0 | 0 |
| 1 | 0 | 0 | 0 | 0 |
| 0 | 1 | 1 | 0 | 0 |
| 0 | 1 | 0 | 0 | 0 |
| 0 | 0 | 1 | 0 | 0 |
| 0 | 0 | 0 | 0 | 0 |

equivalent

## Lesson 12.4

1. Not equivalent; both $x$ and 3 must be multiplied by 2.

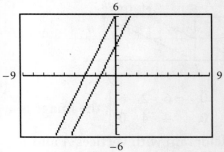

2. Not equivalent; the Distributive Property should be applied. That would give four terms that simplify to two terms, $x^2$ and $-4$.

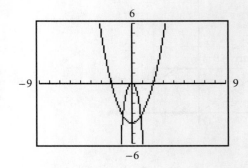

# Answers

3. Not equivalent; multiplication and addition of integers was done incorrectly; $2x - 3x = -1x$ and $2(-3) = -6$.

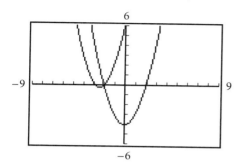

4. Not equivalent; $(2x)^2 = 2^2x^2$, not $2x^2$, which gives a product of $12x^3$.

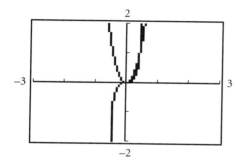

5. The display below shows two real roots. Jenna misread the value of $c$ as 6 rather than $-6$. The roots are 3 and $-2$.

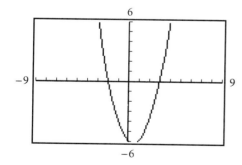